Cognitive Systems Monographs

Volume 19

For further volumes:
http://www.springer.com/series/8354

Beatriz León · Antonio Morales
Joaquín Sancho-Bru

From Robot to Human Grasping Simulation

Springer

Beatriz León
Antonio Morales
Department of Computer Science
and Engineering
Robotic Intelligence Lab
Universitat Jaume I
Castellon
Spain

Joaquín Sancho-Bru
Department of Mechanical Engineering
and Construction
Biomechanics and Ergonomics Group
Universitat Jaume I
Castellon
Spain

ISSN 1867-4925 ISSN 1867-4933 (electronic)
ISBN 978-3-319-34931-2 ISBN 978-3-319-01833-1 (eBook)
DOI 10.1007/978-3-319-01833-1
Springer Cham Heidelberg New York Dordrecht London

Printed on acid-free paper

Springer is part of Springer Science+Business Media (www.springer.com)

Preface

The human hand and its dexterity in grasping and manipulating objects are some of the hallmarks of the human species. For years, anatomic and biomechanical studies have deepened the understanding of the human hands' functioning and, in parallel, the robotics community has been working on the design of robotic hands capable of manipulating objects with a performance similar to that of the human hand. However, although many researchers have partially studied various aspects, to date there has been no comprehensive characterization of the human hands function for grasping and manipulation of everyday life objects.

Our hypothesis is that the confluence of both scientific fields, the biomechanical study of the human hand and the analysis of robotic manipulation of objects, greatly benefits and advances both disciplines. Additionally, we believe that the use of a simulation framework in which we could model and validate each of the processes involved in dexterous grasping is crucially important. Therefore, in this book, the current knowledge of robotics and biomechanics guides the design and implementation of a simulation framework focused on manipulation interactions that allows the study of the grasp through simulation.

In the first part of this book, we detail *OpenGRASP*, a simulation engine focused on robot manipulation interactions embedded in a real grasping cognitive system. Tactile sensor simulation is studied in detail, resulting in a new tactile sensor model. Several applications of the simulator in robot grasping are presented, demonstrating how grasp simulation is a key tool for constructing a world model and understanding the robots environment. Additionally, we demonstrate how to achieve a complete dynamic simulation of a humanoid robot.

In the second part, we use the knowledge acquired from robot simulation to create *OpenHand*, a simulation engine that provides a more comprehensive model of the human hand focused on object grasping and manipulation. It provides a realistic biomechanical hand model of the skeleton, muscles and tendons, including the simulation of the skin and the neuromuscular control. Additionally, it includes tools for grasp analysis such as mechanical contact modelling and control algorithms for closing the hand. We show an application of how the simulation can be used to solve the indeterminate problem of finding the muscular forces that ensures the equilibrium of the grasped object, by minimising different objective functions. Moreover, we propose different quality measures to evaluate various aspects of the human grasps by adapting existing robotic metrics and proposing new measures that consider its biomechanical characteristics, such as muscular fatigue. Finally, the knowledge acquired from the evaluation of grasping in humans is compared with grasping performed by a prosthetic hand demonstrating how the gap between robot and human grasp manipulation could be reduced.

As a result, a valuable framework for the study of the grasp, with relevant applications in several fields such as robotics, biomechanics, ergonomics, rehabilitation and medicine, has been made available to these communities.

Castellón de la Plana
July 2013

Beatriz León
Antonio Morales
Joaquín Sancho-Bru

Contents

Part II Human Grasping Simulation

Chapter 1
Introduction

The human hand and its dexterity in grasping and manipulating objects are some of the hallmarks of the human species. Most of human mechanical interactions with the surrounding world are performed by the hands. We use our hands to perform very different tasks; from exerting high forces (e.g. using a hammer or helping each other carry heavy things) to executing very precise movements (e.g. cutting with a surgical tool or playing an instrument). We also use them to express our feelings, utilising them as a dominant part of our body language. This versatility is possible because of a very complex constitution: a great number of bones connected through different joints, a complicated musculature and a dense nervous system. This complexity is already evident from the kinematics point of view, with more than 23 degrees of freedom [1] controlled by muscles, tendons and ligaments.

For years, anatomic and biomechanical studies have deepened the understanding of their structure, functioning and limitations. There has been an extensive scientific contribution describing and modelling individual components of the human hand: its mechanical structure, the muscular function, the nerve network, mechanical and sensory properties of skin, cognitive functions of manipulation, and many other aspects. However, although many researchers have partially studied various aspects of the human hand from the neuro-physiological and biomechanical viewpoints, to date there has been no comprehensive characterization of the human hand function for grasping and manipulating of everyday life objects, mainly because of the lack of a sufficiently detailed and accurate tool for its simulation.

In parallel, in the field of robotics, the increasing demand for robotic applications in dynamic and unstructured environments and novel situations is motivating the need for dexterous robot hands and grasping abilities, which can cope with the wide variety of tasks and objects encountered in such environments. The scientific community has been working extensively on the design and construction of robotic hands and in all aspects of their control. The explicit goal of these studies is to endow robots with hands capable of manipulating objects with a performance similar to that of the human hand. Although the state of the art in robotics is still far from achieving this purpose, an important body of theoretical and practical knowledge on manipulation

B. León et al., *From Robot to Human Grasping Simulation*,
Cognitive Systems Monographs 19, DOI: 10.1007/978-3-319-01833-1_1,

has been created. Of particular relevance are the advances in the mechanical and mathematical modelling of the interactions between objects, which is a fundamental aspect in the analysis of robotic manipulation.

Our hypothesis is that the confluence of both scientific fields, the biomechanical study of the human hand and the analysis of robotic manipulation of objects, can be a breakthrough in the development of both disciplines. On one hand, the biomechanical study of the human hand is enriched with mathematical analysis techniques to manipulate objects, and on the other the design of robotic hands would benefit from a better understanding of the functioning of the human hand.

1.1 The Grasping Process

Grasping is a core cognitive capability and has been considered as one of the key factors of the evolution of the human brain [2]. In the grasping and manipulation process we can distinguish three main phases: approaching, grasping and manipulating the object. During the approaching phase, the arm moves towards the object and the fingers perform an initial opening to accommodate the object. Subsequently, during the grasping phase, the fingers close to make contact with the object and press it and then the arm and wrist perform the corresponding movements to overcome external forces or torques during the grasp, producing small adjustments in the position of the fingers. Finally, during the manipulation phase, if the task requires changes in the contact points for manipulating the object, coordinated changes are performed in the finger's pressure or position in order to perform such a task without losing the stability of the object. The use of the senses of sight and touch—coordinated by the central nervous system—in the entire process, is critical to the grasp success and efficiency.

Recent works in the grasping literature have different approaches depending on the objectives of the particular study or the researchers' background. Thus, there are papers that focus on analyzing the gripping process from the neurophysiological point of view [3–6], others from the point of view of biomechanics [7–10], while others seek to transfer ideas for robotic manipulators' grip [11–13], some explore the use of other tools [14] and others from the clinical point of view [15, 16].

1.1.1 Grasping in Biomechanics

Mathematical representations, known as biomechanical models of the hand, are used in order to perform qualitative or quantitative analyses on this complex reality. In biomechanics, their use allows studying problems that cannot be analysed directly on humans or that have an experimental cost that is too high; e.g., the study of new alternatives for restoring hand pathologies. Biomechanical models are a description of the hand as a mechanical device: the different elements of the hand are defined in terms of rigid bodies, joints and actuators, and the mechanical laws are applied.

As they are simplified mathematical models of the reality, their use and validity depends on the simplifications considered.

The first biomechanical models of the hand were developed to explain and clarify the functionality of different anatomical elements. In this regard, we can find many works that studied the function of the intrinsic muscles [17–22], others that tried to give an insight into the movement coordination of the interphalangeal joints [23, 24] or studying the causes and effects of different pathologies of the hand [20, 21, 25]. All these models were, however, very limited two-dimensional models allowing only the study of flexion-extension movements, they modelled only one finger, and they included important simplifications. By the year 2000, few three-dimensional models had been developed [26, 27], and none of them modelled the complete hand.

Since 2000, many three-dimensional biomechanical models can be found in literature, having been developed for very different purposes [28–43]: to understand the role of the different anatomical elements, to understand the causes and effects of pathologies, to simulate neuromuscular abnormalities, to plan rehabilitation, to simulate tendon transfer and joint replacement surgeries, to analyse the energetics of human movement and athletic performance, to design prosthetics and biomedical implants, to design functional electric stimulation controllers, to name a few.

All the effort in biomechanics has been focused on appropriately modelling the different hand components (kinematics, muscles, tendons, etc.). Little effort has been spent on the formulation of the grasping problem when using a biomechanical model. In this sense, many limitations persist. Current models do not allow the estimation of the contact information required to use biomechanical models for simulating the grasping of objects. Forces and zones of contact still need to be measured experimentally and input to the model.

In contrast, much research has been carried out on animation techniques over the past years, mainly for use in developing computer games. Lately, these advances have been cleverly used by some ergonomics researchers to develop improved graphical and kinematics hand models for evaluating the use of products [44–47], with good results.

1.1.2 Grasping in Robotics

The phases of human grasping have their counterparts in robotics, although differences in the manipulators dexterity and the available sensory information (visual or tactile) give robotic grasping its own particularities. Grasp planning for robotic multi-finger hands, as well as their dexterous manipulation, are challenging areas of research because of the difficulty posed by the high dimensionality of the configuration space of these manipulators. Besides the hand's internal degrees of freedom, the six degrees of freedom of the arm should also be considered. These describe the position and relative orientation between the object and the robot hand. In fact, despite the effort made over the past decades, there are no robotic hands that can fully emulate the kinematics of the human hand. One of the reasons is the 'hardware': human hands have five soft fingers with high dexterity and compliance compared to less dexterous robot hands featuring very simple contact surfaces. Apart from the hands, compared to humans, robotic arms also have less dexterity and flexibility. For these reasons and also because of the difficulty involved in grasp planning, researchers use robotic hands with few degrees of freedom and employing few predefined grasp preshapes [48].

Miller et al. [49] have predefined grasp preshapes for the Barrett hand with four degrees of freedom and have used primitives to generate the starting positions and directions of the robot hand in planning the grasps. Also, grasp quality measures have also been established [50] to objectively select the best among a predetermined set of grasps. Other researchers [51] have measured the quality of the orientation for the manipulation task to qualify the precomputed grasp.

Another important area of grasp research is focused on finding the location of contact points on the surface of the objects to maximize a certain grasp metric [52, 53]; however, it is difficult to adjust these contact points to feasible manipulator's configurations that are reachable by the robot and avoiding collisions with the environment.

Regarding grasp planning, traditional approaches for controlling robotic hands are typically discarded because they consider each finger as a separate kinematic chain. In these techniques, knowing the object to be grasped and the hand's initial position, the controller synthesizes the desired trajectory of each phalanx in order to complete the grip. Despite the difficulty of this type of control approach, this method was used to control the hand Utah-MIT [54] that is one of the most complex robotic hands that has been produced, with 16 degrees of freedom (4 fingers with 4 finger joints, with tendons emulated by wires) and a separate control for each phalanx.

There are different studies that have tried to plan robotic grasps through imitation or learning by demonstration [55–57]. It is difficult to apply algorithms to learn and imitate the movements of the human hand and translate them to the robotic grasp mainly because robot manipulators have very different kinematics and sensing capabilities with respect to the human hand. In general, motion planning for grasping is performed by simplifying the descriptions of the hand, the object and the task. When using neural networks and knowledge-based systems, a precise representation of the fingers' kinematics is not required [58].

In summary, the planning of robotic grasp for multi-finger hands has been focused on the study of the physical properties of a given grasp or the computation of grasps that meet certain desirable properties. These algorithms use different metrics to determine the quality of the grasps, unlike what happens in the field of the grasping in biomechanics. However, these approaches have failed to deliver practical implementations for a number of reasons, the most crucial being that the methods mostly rely on assumptions that are not satisfied in complex environments with a high degree of uncertainty. Additionally, the grasps studied in robotics are dependent on the characteristics of the robotic hands, and very limited compared with the human hand. Some studies use the learning from demonstration to track the human hand movements, and in the case of using human hand models, they suffer from realism in so far as they greatly simplify its degrees of freedom.

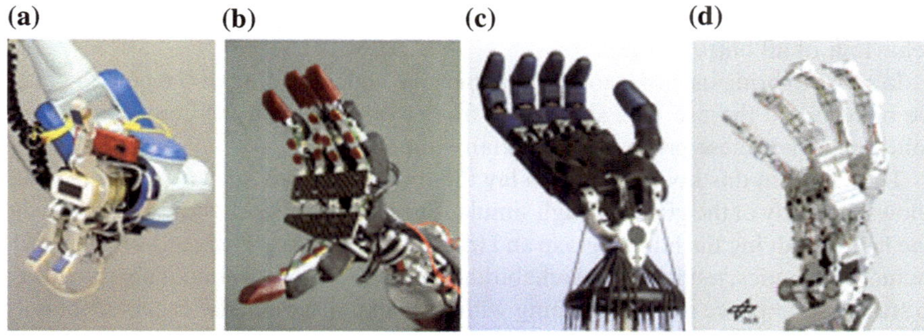

Fig. 1.1 Anthropomorphic robot hands: **a** Barrett Hand (courtesy of the UJI Robotics Intelligent Lab); **b** ARMAR III Hand (courtesy of the Institute for Anthropomatics at KIT); **c** Shadow Hand C5 (courtesy of Shadow Robot Company); **d** Anthropomorphic DLR Hand Arm System (courtesy of DLR Robotics and Mechatronics Centre)

In light of the proliferation of robotic hands (Fig. 1.1) that are becoming increasingly complex and similar to the human hand [59], the field of robotic planning will increasingly be associated with the study of human handling, especially in the area of service robotics where the robot should move in a human environment and manipulate common objects of daily living.

1.2 Simulation: A Tool Towards Understanding the Grasping Process

As we have seen, even with the recent advances in biomechanics and robotics, we have not been able to fully understand or, and to an even lesser degree, able to replicate the process carried out by humans that combines in a natural way: perception, action and predictive capabilities of achieving, mentally planning and then executing a grasp.

There are several questions that still need to be answered, for example: How is the human cognitive system able to plan and control complex manipulation tasks? How does it choose an specific grasp between an infinite set to perform a specific task for a given object? How does the central nervous system select and control the necessary strength required in each muscle to perform the chosen grasp and to counteract the external forces? How are we, from infancy, able to use our experience to learn and refine our capacity for grasping and manipulation?

The difficulty lies in the fact that there are several complex processes and systems involved which interact when humans perform dexterous grasping. First, there is the complex human cognitive system that processes sensory information and controls muscle activity. Second, we have a manipulator as complex as the human hand and arm, which has a highly sensitive sense of touch through the skin, deformable and compliant fingers able to produce soft contacts, more than 20 degrees of freedom, and a complex system of muscles, tendons and ligaments able to move them.

Finally, an environment governed by complex physical laws that characterize the behaviour of all objects within it.

In order to bring us closer to a comprehensive understanding of the human grasp, we believe that the use of a simulation framework in which we could model and validate each of these processes is crucially important.

Therefore, in this book we aim to lay the foundations for such a framework that allows the study of the grasp through simulation. This framework should allow, on the one hand, studying the human grasp and transferring this knowledge to several fields including robotics, medicine and rehabilitation. Furthermore, it should also enable us to study and improve robotic grasping with a system that would be capable, not only of replacing the real hardware, but, more importantly, of being part of the robot's cognitive system, acting as a prediction engine able to emulate the retrieval of self knowledge.

1.3 From Robot to Human Grasping Simulation

In this book, the current knowledge of robotics and biomechanics is used to draw out the rules for developing a simulation framework focused on manipulation interactions that provides the scientific community with a modular, flexible, and accurate tool. The validation and usefulness of the developed framework are shown through a wide set of practical applications.

The objective is two-fold: create a simulation framework that enables us (i) to predict the consequences of a robot grasp and to measure the performance after the execution and (ii) to evaluate the human grasp to achieve a better understanding of the human hand which will help us to transfer this knowledge to the robot's field. *Framework for Grasping Simulation* consists of two parts:

1. **OpenGRASP: A Framework for Robot Grasping Simulation**. A simulation engine focused on robot manipulation interactions embedded in a real grasping cognitive system. This simulation engine enables the system to store and manipulate abstract and specific representations of the perceived objects; to consider possible actions; to predict the results and to plan sequences of actions that complete a desired goal accordingly; to make hypotheses about the real properties of the world; and, finally, to react when an unexpected behaviour occurs. It provides the possibility to develop a grasp reasoning engine including an introspection framework, making its role central in the understanding of human activities and planning new manipulation actions.
2. **OpenHand: A Framework for Human Grasping Simulation**. A simulation engine aiming to obtain a more comprehensive model of the human hand focused on object grasping and manipulation, integrating knowledge and developments from the fields of biomechanics, ergonomics, robotics, and computer animation. It provides a realistic biomechanical hand model of the skeleton, muscles and tendons, including the simulation of the skin and the neuromuscular control.

Additionally, it includes tools for grasp analysis such as mechanical contact modelling, control algorithms for closing the hand and development and application of grasp quality metrics.

The research presented in this book aims to advance the understanding of robot and human grasp, producing a broad set of potential applications of its results. From the society's point of view, it can enhance the capabilities of surgeons for surgical planning in hand operations and open ways to obtaining better bionic limbs that can improve the lives of amputees and the disabled. From the economic point of view, these lines of work could have important medium to long-term implications by opening more possibilities in the field of service robotics to improve grasping capabilities thanks to a better understanding of the hand's functioning, which could result in the development of companies in this sector. From the industrial point of view improving multiple-finger manipulators can enable automation of industrial processes assembly currently being performed manually.

The work proposed herein should be considered as the initial stage in the development of a comprehensive framework for grasping simulation that in the future would consolidate the results obtained in different areas of multidisciplinary knowledge. Therefore a dormant target, but no less important, will be to identify gaps that still exist in the scientific literature for building more challenging future proposals.

1.4 Outline

The book is divided in two main parts. The first one, composed of Chaps. 2 to 4, is devoted to explaining the contributions of robot grasping simulation, while the second, Chaps. 5 to 7, details the evaluation of human grasping using simulation. More precisely, the book is structured as follows.

1.4.1 Part I: Robot Grasping Simulation

Chapter 2: *Robot Grasping Foundations* Introduces the foundation of this book, presenting the basic concepts and definitions involved in the study of object grasping and manipulation tackled in the following chapters.

Chapter 3: *Robot Grasping Simulation* The chapter introduces OpenGRASP, the developed simulation toolkit for grasping and dexterous manipulation, presenting its different components. Tactile sensor simulation is studied in detail, proposing a new tactile sensor model which utilizes collision detection and response methods using soft contacts as well as a full friction description.

Chapter 4: *Applications of Robot Grasping Simulation* In this chapter, the applications of the developed simulator in robot grasping are presented. It is demonstrated how grasp simulation is a key tool for constructing a world model and understanding the robot's environment. Additionally, we demonstrate how to achieve a complete dynamic simulation of a humanoid robot using the developed toolkit.

1.4.2 Part II: Human Grasping Simulation

Chapter 5: *Human Hand Model* This chapter presents a review of the literature regarding biomechanical and ergonomics hand models. The current knowledge on hand models is used to draw out the rules for developing a realistic and self-contained biomechanical model of the hand with special emphasis on grasp and object manipulation. The proposed model consists of a scalable biomechanical model of the human hand composed of bones, tendons, muscles and skin. On this basis, we added closure algorithms to grasp virtual objects, contact models, which allow estimating the transmission of forces in the contacts, and quality indices to provide grasp evaluation tools.

Chapter 6: *Human Grasp Evaluation* This chapter presents a review of the grasp quality measures that have been proposed and then the adaptation of the most common robotic grasp quality measures to the human hand grasp evaluation. Additionally, it presents the proposal of complementary quality indices that may consider biomechanical aspects not taken into account by the robotic indices.

Chapter 7: *Human Grasping Simulation* This chapter presents a study of the adapted grasp quality measures presented in the previous chapter to find the minimum set of indices that enable the evaluation of the different aspects of the human grasp on simulation. Moreover, we present a proposal to calculate a global grasp quality index combining the independent grasp aspects. Finally, the framework for grasp evaluation is used to compare the grasp capabilities of a prosthetic hand with the ones obtained with our human hand model.

References

1. Brand, P., Hollister, A.: Clinical mechanics of the hand. Elsevier Science Health Science div (1992)
2. Napier, J.R., Tuttle, R.: Hands. Princeton University Press, Princeton (1993), http://www.amazon.com/Hands-John-Napier/dp/0691025479
3. Santello, M., Flanders, M., Soechting, J.F.: Postural hand synergies for tool use. J. Neurosci.: Offi. J. Soc., Neurosci. **18**(23), 10105–10115 (1998)
4. Johansson, R.S., Westling, G., Backstrom, A., Flanagan, J.R.: Eye-hand coordination in object manipulation. J. Neurosci.: Off. J. Soc., Neurosci. **21**(17), 6917–6932 (2001)
5. Gentilucci, M., Caselli, L., Secchi, C.: Finger control in the tripod grasp. Experimental brain research. Experimentelle Hirnforschung. Experimentation Cerebrale **149**(3), 351–360 (2003)
6. Winges, S.A., Kornatz, K.W., Santello, M.: Common input to motor units of intrinsic and extrinsic hand muscles during two-digit object hold. J. Neurophysiol. **99**(3), 1119–1126 (2008)
7. Zatsiorsky, V.M., Gao, F., Latash, M.L.: Finger force vectors in multi-finger prehension. J. Biomech. **36**(11), 1745–1749 (2003)

8. Zatsiorsky, V.M., Gao, F., Latash, M.L.: J. Neurophysiol. **95**(4), 2513–2529 (2006)
9. Budgeon, M.K., Latash, M.L., Zatsiorsky, V.M.: Digit force adjustments during finger addition/removal in multi-digit prehension. Experimental Brain Research **189**(3), 345 (2008)
10. Domalain, M., Vigouroux, L., Danion, F., Sevrez, V., Berton, E.: Effect of object width on precision grip force and finger posture. Ergonomics **51**(9), 1441–1453 (2008)
11. Kang, S.B., Ikeuchi, K.: Toward automatic robot instruction from perception—mapping human grasps to manipulator grasps. IEEE Trans. Robot. Autom. **13**(1), 81 (1997)
12. Kaneko, M., Shirai, T., Tsuji, T.: Scale-dependent grasp. IEEE Trans. Syst., Man Cybern. Part A. Syst. Humans **30**(6), 806 (2000)
13. Miller, A.T., Allen, P.K., Santos, V., Valero-Cuevas, F.J.: From robotic hands to human hands: a visualization and simulation engine for grasping research. Indus. Rob. **32**(1), 55 (2005)
14. Armstrong, T.J., Chaffin, D.B.: An investigation of the relationship between displacements of the finger and wrist joints and the extrinsic finger flexor tendons. J. Biomech. **11**(3), 119 (1978)
15. Valero-Cuevas, F.J., Johanson, M.E., Towles, J.D.: Towards a realistic biomechanical model of the thumb: the choice of kinematic description may be more critical than the solution method or the variability/uncertainty of musculoskeletal parameters. J. Biomech. **36**(7), 1019–1030 (2003)
16. Hammond, E.R.A., Shay, B.L., Szturm, T.: Objective evaluation of fine motor manipulation-a new clinical tool. J. Hand Ther.: Off. J. Am. Soc. Hand Ther. **22**(1), 28–35; quiz 36 (2009)
17. Leijnse, J., Kalker, J.J.: A 2-dimensional kinematic model of the lumbrical in the human finger. J. Biomech. **28**(3), 237–249 (1995)
18. Spoor, C.W.: Balancing a force on the fingertip of a two-dimensional finger model without intrinsic muscles. J. Biomech. **16**(7), 497–504 (1983)
19. Spoor, C.W., Landsmeer, J.: Analysis of zigzag movement of human finger under influence of extensor digitorum tendon and deep flexor tendon. J. Biomech. **9**(9), 561–566 (1976)
20. Storace, A., Wolf, B.: Functional-analysis of the role of the finger tendons. J. Biomech. **12**(8), 575–578 (1979)
21. Storace, A., Wolf, B.: Kinematic analysis of the role of the finger tendons. J. Biomech. **15**(5), 391–393 (1982)
22. Thomas, D.H., Long, C., Landsmeer, J.M.F.: Biomechanical consideration of lumbricalis behaviour in the human finger. J. Biomech. **1**, 107–115 (1968)
23. Buchner, H.J., Hines, M.J., Hemami, H.: A dynamic-model for finger interphalangeal coordination. J. Biomech. **21**(6), 459–468 (1988)
24. Lee, J.W., Rim, K.: Maximum finger force prediction using a planar simulation of the middle finger. Proc. Instn. Mech. Eng. Part H: J. Eng. Med. **204**, 169–178 (1990)
25. Smith, E.M., Pearson, J.R., Juvinall, R.C., Bender, L.F.: Role of finger flexors in rheumatoid deformities of metacarpophalangeal joints. Arthritis Rheum **7**(5P1), 467 (1964)
26. Biryukova, E.V., Yourovskaya, V.Z.: A model of human hand dynamics, pp. 107–122. Advances in the Biomechanics of the Hand and Wrist, Plenum Press (1994)
27. Casolo, F., Lorenzi, V.: Finger mathematical modelling and rehabilitation, pp. 197–223. Advances in the Biomechanics of the Hand and Wrist, Plenum Press (1994)
28. Fok, K.S., Chou, S.M.: Development of a finger biomechanical model and its considerations. J. Biomech. **43**(4), 701–713 (2010)
29. Kamper, D., Fischer, H., Cruz, E.: Impact of finger posture on mapping from muscle activation to joint torque. Clin. Biomech. **21**(4), 361–369 (2006)
30. Kubus, D., Iser, R., Winkelbach, S., Wahl, F.M.: Efficient parallel random sample matching for pose estimation, localization, and related problems. In: Kröger, T., Wahl, F.M. (eds.) Advances in Robotics Research, pp. 239–250. Springer Berlin Heidelberg (2009)
31. Lee, S.W., Chen, H., Towles, J.D., Kamper, D.G.: Estimation of the effective static moment arms of the tendons in the index finger extensor mechanism. J. Biomech. **41**(7), 1567–1573 (2008)
32. Lee, K.S., Mo, S.M., Hwang, J.J., Wang, H., Jung, M.C.: Relaxed hand postures. Japan. J. Ergonomics **44**(Supplement), 436–439 (2008)

33. Qiu, D., Fischer, H.C., Kamper, D.G.: Muscle Activation Patterns during Force Generation of the Index Finger. In: Engineering in: Medicine and Biology Society, 2009. EMBC 2009. Annual International Conference of the IEEE, pp. 3987-3990, 2009
34. Roloff, I., Schoffl, V., Vigouroux, L., Quaine, F.: Biomechanical model for the determination of the forces acting on the finger pulley system. J. Biomech. **39**(5), 915–923 (2006)
35. Sancho-Bru, J.L., Perez-Gonzalez, A., Vergara-Monedero, M., Giurintano, D.: A 3-d dynamic model of human finger for studying free movements. J. Biomech. **34**(11), 1491–1500 (2001)
36. Sancho-Bru, J.L., Giurintano, D.J., Pérez-González, A., Vergara, M.: Optimum tool handle diameter for a cylinder grip. J. Hand Ther.: Off. J. Am. Soc. Hand Ther. **16**(4), 337–342 (2003)
37. Sancho-Bru, J.L., Perez-Gonzalez, A., Vergara, M., Giurintano, D.J.: A 3d biomechanical model of the hand for power grip. J. Biomech. Eng. **125**(1), 78–83 (2003)
38. Sancho-Bru, J., Vergara, M., Rodríguez-Cervantes, P.J., Giurintano, D., Pérez-González, A.: Scalability of the muscular action in a parametric 3d model of the index finger. Ann. Biomed. Eng. **36**, 102–107 (2008)
39. Valero-Cuevas, F.J.: Predictive modulation of muscle coordination pattern magnitude scales fingertip force magnitude over the voluntary range. J. Neurophysiol. **83**(3), 1469–1479 (2000)
40. Valero-Cuevas, F.J.: An integrative approach to the biomechanical function and neuromuscular control of the fingers. J. Biomech. **38**(4), 673–684 (2005)
41. Vigouroux, L., Quaine, F., Labarre-Vila, A., Moutet, F.: Estimation of finger muscle tendon tensions and pulley forces during specific sport-climbing grip techniques. J. Biomech. **39**(14), 2583–2592 (2006)
42. Vigouroux, L., Ferry, M., Colloud, F., Paclet, F., Cahouet, V., Quaine, F.: Is the principle of minimization of secondary moments validated during various fingertip force production conditions? Human Mov. Sci. 27(3), 396–407 (2008)
43. Wu, J.Z., An, K.N., Cutlip, R.G., Dong, R.G.: A practical biomechanical model of the index finger simulating the kinematics of the muscle/tendon excursions. Bio-Med. Mater. Eng. **20**(2), 89–97 (2010)
44. Endo, Y., Kanai, S., Kishinami, T., Miyata, N., Kouchi, M., Mochimaru, M.: Virtual grasping assessment using 3d digital hand model. In: 10th Annual Applied Ergonomics Conference: Celebrating the Past: Shaping the Future (12 March 2007 through 15 March 2007)
45. Endo, Y., Kanai, S., Miyata, N., Kouchi, M., Mochimaru, M., Konno, J., Ogasawara, M., Shimokawa, M.: Optimization-based grasp posture generation method of digital hand for virtual ergonomics assessment. SAE Int. J. Passeng. Cars - Electron. Electr. Syst. **1**(1), 590–598 (2008)
46. Goussous, F.A.: Grasp planning for digital humans. Ph.D. thesis, Iowa University (2007)
47. Kawaguchi, K.: Database-driven grasp synthesis and ergonomic assessment for handheld product design. Lect. Notes Comput. Sci. **5620**, 642–652 (2009)
48. Wren, D., Fisher, R.: Dextrous hand grasping strategies using preshapes and digit trajectories. In: IEEE International Conference on Systems, Man and, Cybernetics. vol. 1, pp. 910–915, 1995
49. Miller, A.T., Knoop, S., Christensen, H., Allen, P.K.: Automatic grasp planning using shape primitives. In: Robotics and Automation, 2003. Proceedings. ICRA'03. IEEE International Conference on. vol. 2, pp. 1824–1829. IEEE, 2003
50. Bicchi, A.: On the closure properties of Robotic grasping. Int. J. Rob. Res. **14**, 319–334 (1995)
51. Borst, C., Fischer, M., Hirzinger, G.: Grasp planning: how to choose a suitable task wrench space. IEEE, 2004.
52. Li, Z., Sastry, S.: Task-oriented optimal grasping by multifingered robot hands. IEEE J. Rob. Autom. **4**(1), 32–44 (1987)
53. Zhu, X., Wang, J.: Synthesis of force-closure grasps on 3-d objects based on the q distance. IEEE Transactions on Robotics 19(4), 669–679 (2003), http://dblp.uni-trier.de/db/journals/trob/trob19.html#ZhuW03
54. Jacobsen, S.C., Iversen, E.K., Knutti, D.F., Johnson, R.T., Biggers, K.B.: Design of the Utah/MIT dextrous hand. In: Robotics and Automation. Proceedings. IEEE International Conference on 1986, vol. 3, pp. 1520–1532, IEEE, 1986

55. Aleotti, A., Caselli, S.: Grasp recognition in virtual reality for robot pre grasp planning by demonstration. In: Proceedings - IEEE International Conference on Robotics and Automation, p. 2801, 2006
56. Romero, J., Kjellstrm, H., Kragic, D.: Human-to-robot mapping of grasps. In: Proceedings of IEEE/RSJ International Conference on Intelligent Robots and Systems, WS on Grasp and Task Learning by Imitation, 2008.
57. Harada, K., Kaneko, K., Kanehiro, F.: Fast grasp planning for hand/arm systems based on convex model. Proceedings - IEEE International Conference on Robotics and Automation, p. 1162, 2008.
58. Molina-Vilaplana, J., López-Coronado, J.: Neural modelling of hand grip formation during reach to grasp. Neurocomputing **71**(1–3), 411 (2007)
59. Parada, J.E., Nava, N.E., Ceccarelli, M.: A Methodology for the Design of Robotic Hands with Multiple Fingers. Int. J. Adv. Rob. Syst. **5**(2), 177–184 (2008)

Part I
Robot Grasping Simulation

Chapter 2
Robot Grasping Foundations

2.1 Introduction

In this book, we focus on the grasping problem, consisting of determining the grasp required to carry out certain manipulation tasks on an object.

Definition 2.1 A *grasp* is commonly defined as a set of contacts on the surface of the object, which purpose is to constrain the potential movements of the object in the event of external disturbances [1–4].

For a specific robotic hand, different grasp types are planned and analysed in order to decide which one to execute. A contact model should be defined to determine the forces or torques that the robot manipulator must exert on the contact areas. Most of the work in robotics assume point contacts, and larger areas of contact are usually discretized to follow this assumption [2]. Two main problems can be distinguished in robotic grasping: analysis and synthesis [5].

Definition 2.2 *Grasp analysis* consists on finding whether the grasp is stable using common closure properties, given an object and a set of contacts. Then, quality measures can be evaluated in order to enable the robot to select the best grasp to execute.

Definition 2.3 *Grasp synthesis* is the problem of finding a suitable set of contacts given an object and some constraints on the allowable contacts.

In the following sections, a detailed description of the contact models and the most common approaches for grasp analysis and synthesis are presented. The definitions of the terminology and notation are mainly taken from [3, 6] where the reader is referred to find a complete overview of the modelling of contact interfaces and an introduction of the fundamental models of grasp analysis.

B. León et al., *From Robot to Human Grasping Simulation*,
Cognitive Systems Monographs 19, DOI: 10.1007/978-3-319-01833-1_2,

Table 2.1 Notations

$\{W\}$	Word coordinate frame
$\{O\}$	Object coordinate frame
n_c	Number of contact points
c_i	Contact point i relative to $\{W\}$
$\{C\}_i$	Contact point i coordinate frame with axis $\{\hat{n}_i, \hat{t}_i, \hat{o}_i\}$
$\hat{n}_i$	Unit normal to the contact tangent plane directed toward the object
p	Position of the object relative to $\{W\}$
v	Linear velocity of point p
ω	Angular velocity of the object relative to $\{W\}$
w_i	Generalized force acting on the object for a unit force along $\hat{n}_i$
f_i	Force applied to the object at the point c_i
τ_i	Resulting moment at point p
w_o	Total set of wrenches that can be transmitted to the object through the n_c
w_{ext}	Disturbing external wrenches
μ	Friction coefficient of the contacting materials
β	Half-angle of the friction cone
m	Number of faces of discretized friction cone
B	Selection matrix
l	Total number of twist components transmitted
G	Grasp matrix
$\tilde{G}_i$	Partial grasp matrix
$\tilde{G}$	Complete grasp matrix
J	Hand Jacobian matrix
$\tilde{J}_i$	Partial hand Jacobian matrix
$\tilde{J}$	Complete hand Jacobian matrix
G_J	Grasp Jacobian matrix

2.2 Contact Modelling

2.2.1 Contact Kinematics

Consider a manipulator contacting a rigid body whose position and orientation is specified by the location of the origin of a coordinate frame $\{O\}$ fixed to the object and the orientation of this coordinate frame relative to an inertial frame $\{W\}$ fixed in the world (see Fig. 2.1). Let $p \in \mathbb{R}^3$ be the position of the object and $c_i \in \mathbb{R}^3$ the location of a contact point i relative to $\{W\}$. At this contact point, we define a frame $\{C\}_i$ with axis $\{\hat{n}_i, \hat{t}_i, \hat{o}_i\}$ where $\hat{n}_i$ is the unit normal to the contact tangent plane and is directed toward the object. The other two unit vectors are orthogonal and lie in the tangent plane of the contact. For readers' convenience, a list of notations is given in Table 2.1.

Definition 2.4 A *twist* is the representation of the spacial velocity of the object and can be written as $t \in \mathbb{R}^6$:

$$t = \begin{pmatrix} \upsilon \\ \omega \end{pmatrix} \tag{2.1}$$

where $\upsilon \in \mathbb{R}^3$ is the linear velocity of point p and $\omega \in \mathbb{R}^3$ is the angular velocity of the object in the world frame $\{W\}$.

Definition 2.5 A *contact* can be defined as a joint between the finger and the object. The shape of the contacting surfaces and the stiffness and frictional characteristics of the contacting bodies define the nature of this joint [5].

A point contact acting on the object provides a unilateral constraint which prevents the object from locally moving against the contact normal [6].

Definition 2.6 The force applied by a finger at the contact point generates a *wrench* on the object with force and torque components, represented by vector $w_i \in \mathbb{R}^6$:

$$w_i = \begin{pmatrix} f_i \\ \tau_i \end{pmatrix} = \begin{pmatrix} \hat{n}_i \\ (c_i - p) \times \hat{n}_i \end{pmatrix} \tag{2.2}$$

where $f_i \in \mathbb{R}^3$ is the force applied to the object at the point c_i and $\tau_i \in \mathbb{R}^3$ the resulting moment at point p.

As forces and torques are dimensionally different, a parameter ρ is introduced:

$$w_i = \begin{pmatrix} f_i \\ \tau_i / \rho \end{pmatrix} \tag{2.3}$$

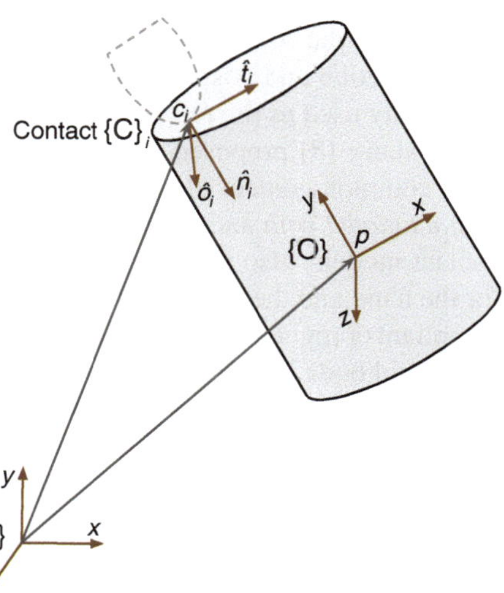

Fig. 2.1 Notation for an object in contact with a manipulator

Selecting ρ to have *length* units, allows all the components of w to have units of *force*. Two approaches to define ρ have been presented in [7]. One approach considers ρ as the largest distance from the object's centre of mass to any point of the object. This restricts the maximum torque to the maximum applied force, which is typically considered unitary. The other approach considers ρ as the radius of gyration of the object. It has a physical meaning in terms of energy but is less used and more complex to calculate.

If there are multiple contacts acting on an object, the total set of wrenches w_o that can be transmitted to the object through the n_c contacts is the linear combination of all individual wrenches:

$$w_o = \sum_{i=1}^{n_c} w_i \tag{2.4}$$

To prevent a grasp from slipping, the forces in the contacts (and their corresponding wrenches w_i) and any disturbing force or wrench w_{ext} have to be in equilibrium:

$$w_o + w_{ext} = 0 \tag{2.5}$$

This equation is valid always that the contact forces satisfy the contact constrains described in the next section.

2.2.2 Contact Models

A contact model maps the forces that can be transmitted through the contact to the resultant wrenches w_i relative to the object. This map is determined by the geometry of the contacting surfaces and the material properties of the objects, which dictate friction and possible contact deformation [6]. The object's centre of mass is commonly used as the reference point p in the object.

Salisbury [8] proposed a taxonomy of eight contact models. Among these, the most common contact models used in robotic grasping (see Fig. 2.2) are the *point contact model with* and *without friction* and the *soft-finger contact model* [9]. Point contact models, also named rigid-body contact models, assume rigid-body models for the hand and the grasped object while the soft-finger contact models, also called compliant or regularised models, assume that the hand is a deformable element grasping a rigid body [6]. The former models assume the collision to be an instantaneous and discontinuous phenomenon (discrete event) and the equations of motion are derived by balancing the system's momenta before and after the impact. In contrast, compliant models describe the normal and tangential compliance relations over time.

Point contact without friction

A point contact without friction can only transmit forces along the normal to the object surface at the contact point. No deformations are allowed at the points of

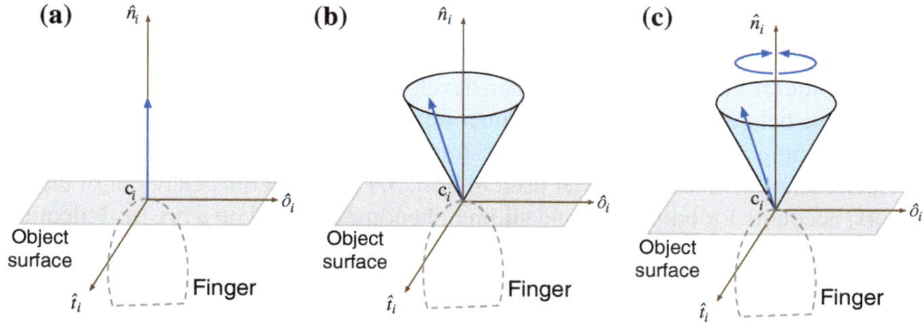

Fig. 2.2 Contact models commonly used in robotics: **a** Contact without friction, **b** Contact with friction and **c** Soft-finger contact

contact between the two bodies and, instead, contact forces arise from the constraint of incompressibility and impenetrability between the rigid bodies. This model is used when the contact patch is very small and the surfaces of the hand and object are slippery [3]. It does not represent the real contact situations that appear in robotic manufacturing operations [10, 11] and, when used, the machine accuracy is negatively affected. Moreover, they are not capable of predicting the individual contact forces of a multiple-contact fixture [12, 13].

Point contact with friction

A point contact with friction is used when there is significant contact friction, but the contact patch is small so that no appreciable friction moment exists [3]. Therefore, it can transmit forces in the normal and tangential directions to the surface at the contact point but non of the moment components are transmitted.

A number of models have been developed which attempt to capture the essence of the complicated friction phenomena [14]. The classical model, called Coulomb friction, is based on the idea that friction opposes motion and that its magnitude is independent of the velocity and contact area. It is an empirical model that asserts that the allowed tangential force is proportional to the applied normal force by $f_t \leq \mu f_n$, where μ is called the friction coefficient of the contacting materials. The friction forces can be represented geometrically as a friction cone where the set of all forces that can be applied is constrained to be inside a cone centred about the surface normal (see Fig. 2.3) with half-angle $\beta = tan^{-1}(\mu)$. In the spacial case, the friction cone is a circular cone, defined by

$$\sqrt{f_{t_i}^2 + f_{o_i}^2} \leq \mu f_{n_i}, \; f_{n_i} \geq 0. \tag{2.6}$$

For computational purposes, the friction cone can be approximated by an inscribed regular polyhedral cone with m faces, as shown in Fig. 2.3b. The largest the m the better the approximation, but also the larger the computational cost. The wrenches generated by forces along the edge of the discretized friction cone are referred to as *primitive wrenches*.

Many other friction models have been developed as aggregate models of complex microscopic behaviour with different functional dependencies on factors such as the speed of sliding and the duration of the static contact before sliding. For example, the LuGre friction model [15] later used in Sect. 3.4 to model the behaviour of tactile sensors, accounts for both static and sliding phenomena based on a bristle deflection interpretation.

In this contact model, contact forces arise from two sources: the rigid-body model assumption for both the hand and the object, and the frictional forces. The use of this contact model in the manipulation planning problem has led to some interesting conclusions. There may be multiple solutions to a particular problem (ambiguity) or there may be no solutions (inconsistency) [16]. This kind of contact models have been used for analysing the tip-over problem of a planar object that is being pushed by a finger from a quasistatic point of view, i.e. neglecting the dynamic properties of the robot and the object [17–19].

Soft-finger contact

Finally, the soft contact model is used when the surface friction and the contact patch are large enough to generate significant friction forces and a friction moment about the contact normal. It is used to model the contact between a soft finger and a rigid object allowing the finger to apply an additional torsional moment with respect to the normal at the contact point [20–25].

In order to model the pressure distribution in the contact area different models have been developed that fall into three main categories: analytical elasticity-based models, elastic foundation models (EFM) and finite element models (FEM) [26]. Analytical models are based on theoretical formulations of elasticity calculating contact areas and stresses on both the surface and the sub-surface of the contacting bodies,

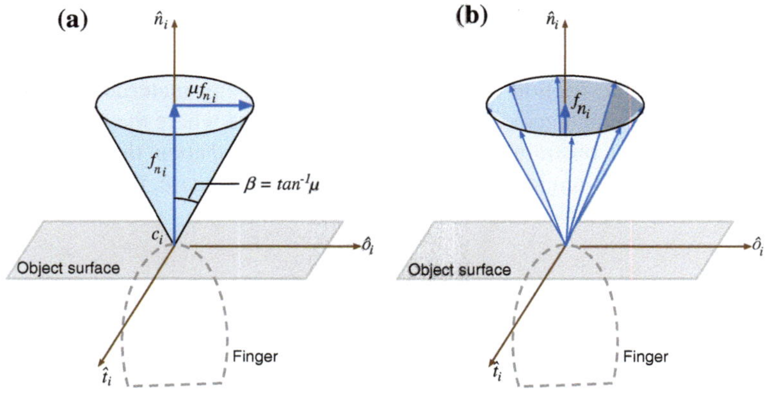

Fig. 2.3 Friction cone: **a** Spacial representation and **b** approximation as an inscribed polyhedral cone

but are restricted to simple geometries. The classical Hertz contact model [27, 28] and others derived from it are part of this category. However, robotic fingertips are made of nonlinear elastic materials. For that reason, the Hertzian contact model does not accurately represent this type of contact. EFMs were developed in order to allow a simple discrete contact calculation in more general surface geometries modelling the deformable part of the contact as a layer over a rigid base, and a series of discrete and independent springs in the contact normal direction. An application of this category of models is presented in Sect. 3.4 where it is used to create a simulation of a general tactile sensor and validated in robot grasping applications. FEMs have been increasingly used over recent years given that they supply information about the sub-surface stresses and strain in volumetric finite elements. However, they are excessively time consuming for fast simulation in dynamic grasping and manipulation models. Therefore, simplified numerical models are interesting alternatives.

These soft contact models have been used in robotic applications. Xydas et al. [29, 30] presented a power-law theory for modelling nonlinear elastic contacts present in robotic fingers. More realistic and complicated models have been developed in the last few years that better represent the contact mechanics for soft fingers [20, 21, 31]. A soft contact model based on that of Ciocarlie et al. (2005) was used in this work to model the contact between objects and the human hand (Sect. 5.5.4).

Despite being the soft contact model the more accurate, it is the hard finger contact with friction the one that is used more often in robotics given the complexity and time-consuming restrictions to simulate real activities.

2.2.3 Selection Matrices

Each one of the previous contact models selects components of the contact twists to transmit between the hand and the object. This is done by equating a subset of the components of the hand and object twist at each contact [3].

Definition 2.7 A particular contact model is defined through the *selection matrix* $B_i \in \mathbb{R}^{l_i \times 6}$, which works like a filter selecting l_i components of the relative contact twist and sets them to zero.

$$B_i(t_{i,hand} - t_{i,obj}) = 0 \tag{2.7}$$

B_i and l_i can be defined for each contact model as:
Point contact without friction

$$l_i = 1, B_i = \begin{bmatrix} 1 & 0 & 0 & 0 & 0 & 0 \end{bmatrix} \tag{2.8}$$

Point contact with friction

$$l_i = 3,\ B_i = \begin{bmatrix} 1 & 0 & 0 & 0 & 0 & 0 \\ 0 & 1 & 0 & 0 & 0 & 0 \\ 0 & 0 & 1 & 0 & 0 & 0 \end{bmatrix} \tag{2.9}$$

Soft-Finger contact

$$l_i = 4,\ B_i = \begin{bmatrix} 1 & 0 & 0 & 0 & 0 & 0 \\ 0 & 1 & 0 & 0 & 0 & 0 \\ 0 & 0 & 1 & 0 & 0 & 0 \\ 0 & 0 & 0 & 1 & 0 & 0 \end{bmatrix}. \tag{2.10}$$

After choosing a particular contact model for each contact, the *selection matrix* B for all n_c contacts can be calculated as:

$$B = Blockdiag(B_1, \ldots, B_{n_c}) \in \mathbb{R}^{l \times 6n_c} \tag{2.11}$$

where the total number of twist components l transmitted through the n_c contacts is given by:

$$l = \sum_{i=1}^{n_c} l_i \tag{2.12}$$

2.3 Grasp Analysis

Once the contact model has been established, it can be used to study tasks involving multiple contacts. The set of contacts defining each grasp can be analysed in order to test the grasp's ability to resist disturbances and its dexterity properties. As it is presented afterwards, the grasps that can be maintained for every possible disturbing load are known as *closure grasps*. However, there is usually more than one grasp that fulfils this condition. That is why many metrics and approaches have been proposed to evaluate the dexterity of the selected grasps and determine which one is the best to be executed.

2.3.1 Grasp Matrix and Hand Jacobian

There are two matrices used in grasp analysis: *the grasp matrix* G and *the hand Jacobian* J. They define the relevant velocity kinematics and force transmission properties of the contacts. They are introduced here, but a complete explanation can be found in [3].

Partial Grasp Matrix The transpose of the *partial grasp matrix* $\tilde{G}_i \in \mathbb{R}^{6\times 6}$ maps the object twist from $\{W\}$ to the contact frame:

$$t_{i,obj} = \tilde{G}_i^T t \tag{2.13}$$

where t and $t_{i,obj}$ denote the object twist related to $\{W\}$ and to $\{C\}_i$, respectively.

$\tilde{G}_i$ can be calculated as:

$$\tilde{G}_i = \begin{pmatrix} R_i & 0 \\ S(c_i - p)R_i & R_i \end{pmatrix} \tag{2.14}$$

where $R_i \in \mathbb{R}^{3\times3}$ represents the rotation matrix of the $\{C\}_i$ contact frame with respect to $\{W\}$, c_i the position of the contact point, p the position of the object and $S(c_i - p)$ is the cross-product matrix, that given a three-vector $r = [r_x, r_y, r_z]^T$, $S(r)$ is defined as:

$$S(r) = \begin{pmatrix} 0 & -r_z & r_y \\ r_z & 0 & -r_x \\ -r_y & r_x & 0 \end{pmatrix} \tag{2.15}$$

Complete Grasp Matrix The *complete grasp matrix* $\tilde{G} \in \mathbb{R}^{6\times6n_c}$ is the combination of the partial grasp matrices for each of the n_c contact points.

$$\tilde{G}^T = \begin{pmatrix} \tilde{G}_1^T \\ \vdots \\ \tilde{G}_{n_c}^T \end{pmatrix} \tag{2.16}$$

It maps the object twist from {W} to {C}:

$$t_{c,obj} = \tilde{G}^T t \tag{2.17}$$

where $t_{c,obj} \in \mathbb{R}^{6n_c}$ is a vector containing all the twist of the object in the contact frames:

$$t_{c,obj} = (t_{1,obj}^T \dots t_{n_c,obj}^T)^T \tag{2.18}$$

Partial Hand Jacobian The partial hand Jacobian $\tilde{J}_i \in \mathbb{R}^{6\times n_q}$ maps the manipulator joint velocities $\dot{q}$ to the contact twists on the hand $t_{i,hand}$, expressed in the contact frame $\{C\}_i$:

$$t_{i,hand} = \tilde{J}_i \dot{q} \tag{2.19}$$

where $q = [q_1 \dots q_{n_q}]^T$ represents the vector of joint displacements and n_q the number of hand joints.

$\tilde{J}_i$ can be calculated as:

$$\tilde{J}_i = R_i \begin{pmatrix} d_{i,1} \dots d_{i,n_q} \\ l_{i,1} \dots l_{i,n_q} \end{pmatrix} \tag{2.20}$$

where:

$$d_{i,j} = \begin{cases} 0_{3\times 1} & \text{if contact i does not affect the joint j} \\ S(c_i - \zeta_j)^T \hat{z}_j & \text{if joint j is revolute} \end{cases} \tag{2.21}$$

$$l_{i,j} = \begin{cases} 0_{3\times 1} & \text{if contact i does not affect the joint j} \\ \hat{z}_j & \text{if joint j is revolute} \end{cases} \tag{2.22}$$

being ζ_j the origin of the coordinate frame associated with the j-th joint and $\hat{z}_j$ the unit vector in the direction of the rotational axis for the revolute joint, expressed in $\{W\}$.

Complete Hand Jacobian $\tilde{J}$ The complete hand Jacobian $\tilde{J}$ $\in \mathbb{R}^{6n_c \times n_q}$ is the combination for each of the n_c contact points:

$$\tilde{J} = \begin{pmatrix} \tilde{J}_1 \\ \vdots \\ \tilde{J}_{n_c} \end{pmatrix} \tag{2.23}$$

It maps the joint velocities to the twists of the hand expressed in the contact frames:

$$t_{c,hand} = \tilde{J}\dot{q} \tag{2.24}$$

where $t_{c,hand} \in \mathbb{R}^{6n_c}$ is a vector containing all the twist of the hand in the contact frames:

$$t_{c,hand} = (t_{1,hand}^T \dots t_{n_c,hand}^T)^T \tag{2.25}$$

Grasp Matrix G and Hand Jacobian J After choosing a transmission model for each contact, as explained in Sect. 2.2.2, the contact constraint equations (Eq. 2.7) for all n_c contacts can be written in compact form as:

$$B(t_{c,hand} - t_{c,obj}) = 0 \tag{2.26}$$

By substituting into this equation $t_{c,hand}$ (Eq. 2.24) and $t_{c,obj}$ (Eq. 2.17) one obtains:

$$B(\tilde{J}\dot{q} - \tilde{G}^T t) = 0 \tag{2.27}$$

Defining the *grasp matrix G* and *hand Jacobian J* as:

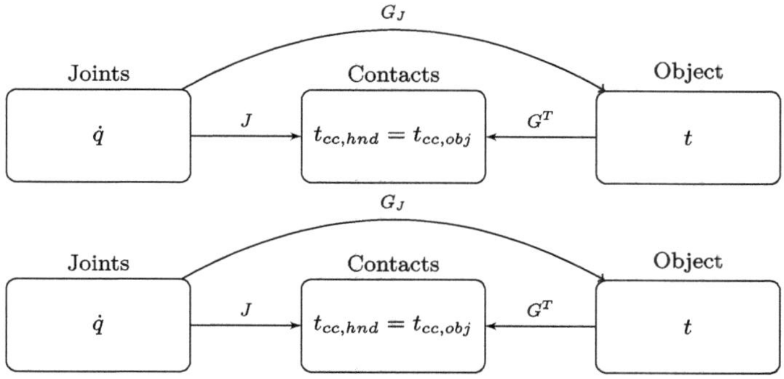

Fig. 2.4 Diagram of relationships between velocities for a multi-fingered grasp

$$G^T = B\tilde{G}^T \tag{2.28}$$

$$J = B\tilde{J}, \tag{2.29}$$

the contact constraint equations for all n_c contacts can be written as:

$$(J - G^T)\begin{pmatrix}\dot{q}\\ t\end{pmatrix} = 0 \tag{2.30}$$

or:

$$J\dot{q} = t_{cc,hnd} = t_{cc,obj} = G^T t \tag{2.31}$$

where $t_{cc,hnd}$ and $t_{cc,obj}$ contain only the components of the twist that are transmitted by the contacts.

Grasp Jacobian The grasp Jacobian G_J is the transformation from the joint velocities to the velocity of the object being grasped [32]:

$$t = G_J\dot{q} \tag{2.32}$$

It takes into account the transformations for each finger from joint velocities to fingertip Cartesian velocity (J), the contact relationships and the transformations from the contact frames of reference to the object frame of reference (G). Thus, it is a function of the hand posture and the lengths of the finger segments.

$$G_J = (G^+)^T J \tag{2.33}$$

with G^+ being the generalized inverse of G. Figure 2.4 summarizes the relationships between velocities in a multi-fingered grasp.

2.3.2 Disturbance Resistance

The first test for evaluating a grasp consists of determining its ability to constrain the motions of the manipulated object and to apply arbitrary contact forces on the object without violating friction constraints at the contacts [33]. Two commonly used properties have been proposed to ensure this condition: force and form closure.

Definition 2.8 A grasp is in *force-closure* if the fingers can apply, through the set of contacts, arbitrary wrenches on the object, which means that any motion of the object is resisted by the contact forces [34].

Definition 2.9 A grasp is in *form-closure* if the location of the contact points on the object ensures its immobility [33].

Form closure is a stronger condition than force closure and it is mostly used when executing power grasps [3]. Force closure is possible with fewer contacts, making it suitable for executing precision grasps, but it requires the ability to control internal forces.

The analysis of form closure is intrinsically geometric. [35] stated that a necessary and sufficient condition for form-closure is that the contact wrenches of the grasp positively span the whole wrench space.

Definition 2.10 A *grasp wrench space* (*GWS*) is the space of wrenches that can be applied to the object at each contact point.

The boundary of the wrench space can be calculated as a convex hull. Form-closure then can be equivalently determined verifying if the origin of the wrench space lies inside this convex hull [36]. Based on the above necessary and sufficient conditions, many tests that have been proposed by [34, 37, 38]. Ferrari and Canny [39] is the most widely-used. They proposed to calculate the radius of the largest ball inscribed in the convex hull centred in the origin and verify that it is larger than zero. Zhu and Wang [40] developed a numerical test which measures the scaling factor for the maximum compact set inscribed in the *GWS* with centre in the origin.

Assessing the force-closure property of a robotic grasp is much more difficult because of the nonlinear nature of the Coulomb friction cone [40]. [41] formulated the force-closure test as 12 nonlinear programming problems. Trinkle [42] formalized the force closure condition as a linear programming problem. [33] observed that the force-closure problem is equivalent to the stability of an ordinary differential equation. [43] reformulated the force closure condition as a ray-shooting problem by linearizing the friction cones and proposed a clean-cut test for force closure grasps. [44] proposed a force-closure test representing the nonlinear friction cone constraints as linear matrix inequalities, for which efficient algorithms are now available. With the linearization of the friction cone, most of the existing form-closure tests can be generalized to force-closure analysis. [45] proposed a numerical criterion for 3-D grasps with frictional point contacts or soft contacts, formulated as a convex constrained optimization problem without linearization of the friction cone. More recently, [46] proposed an algorithm for computing the distance between a point and a convex cone in n-dimensional space that can be applied to force-closure test and improve their efficiency.

2.3.3 *Optimal Contact Forces Computation*

The actual finger forces for a given grasp will be obtained by considering that they have to satisfy the dynamic equilibrium of the grasped object. Since the number of contacts is usually more than necessary, there is not a unique set of forces that ensures the equilibrium. Therefore the problem, referred to as *grasping force optimization* problem, needs to be solved computing the optimal finger forces that satisfy some optimization criterion, such as minimizing the magnitudes or inclination angles of contact forces [47].

It is a constrained optimization problem, for which a variety of general optimization methods are applicable. There have been a number of algorithms proposed in the last two decades aiming to solve it [43, 44, 48–53]. Recently Zheng et al. have proposed new algorithms to improve the computation efficiency since it is desirable to obtain the contact force distribution in real-time [47, 54].

2.4 Grasp Synthesis

Given an object, grasp synthesis algorithms should provide a suitable set of contacts on the object surface and determine an appropriate hand configuration. Usually these algorithms take the geometry of the object as an input to select optimal force-closure contact locations or whole regions that yield force closure. These contacts are the starting point for grasp analysis and dexterous manipulation methods. Some approaches give only information about the finger contact locations on the object without considering the hand constraints. They can result in stable grasps that are not reachable in practice by the robot hand. Moreover, even if they are reachable, it is difficult to position the fingers precisely on the contact points because there will be always unavoidable errors locating the end-effector [55].

A number of force-closure test have been proposed based on specific geometric conditions for a given number of fingers. [34] proposed a geometric method for computing maximal independent two-finger grasps on polygons. [56] proposed an approach for synthesizing a three-fingered grasp on polygonal objects, and later extended to a four-fingered grasp on polyhedral objects [57], based on the concept of independent contact regions. [58] developed an algorithm for calculating all force-closure grasp configurations on polygons using the computational geometry technique. However these approaches are not suitable for the generic problem of planning an optimal force-closure grasps on general three-dimensional objects with any number of contact points.

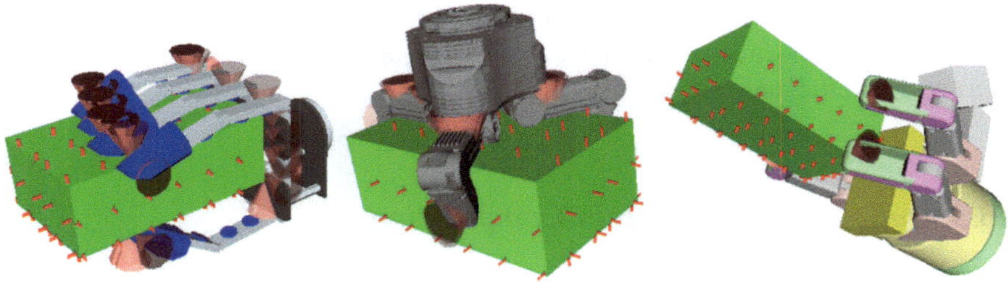

Fig. 2.5 Testing the set of hand preshapes for several robot hands using OpenRAVE

Alternative approaches, called knowledge-based approaches, have considered the configuration of the hand by generating the grasp with a predefined set of hand postures. The idea of hand preshapes started with studies of the human prehension capabilities [59] that introduced the distinction between power and precision grasps. Following this work, [10] created a taxonomy in which details of the task and the object geometry are taken into account. Since then, several papers have adopted this approach for grasping [61, 55, 60]. [62] used the GraspIt! simulator [63] to test the set of hand preshapes on a 3D model of the object. Using a simulator has many advantages, including the ability to plan grasps in complex environments involving obstacles and also to check the reachability constraints of the robot arm. More recently OpenRAVE, a planning architecture that has a more flexible design, has been proposed to automate this process [64]. Examples of the generation of grasps hypotheses for several robot hands can be seen in Fig. 2.5.

Grasp synthesis is presented and used in detail in Chaps. 3 and 4 for different robotic grasp applications.

References

1. Murray, R.N., Li, Z., Sastry, S.: A mathematical introduction to robotics manipulation. CRC Press, USA (1994)
2. Bicchi, A., Kumar, V.: Robotic grasping and contact: a review. In: Robotics and Automation, 2000. Proceedings. ICRA '00. IEEE International Conference on. pp. 348–353. IEEE (2000)
3. Prattichizzo, D., Trinkle, J.: Springer Handbook of Robotics, pp. 671–700. Springer, Berlin (2008)
4. Roa Garzón, M.: Grasp planning methodology for 3D arbitrary shaped objects. Ph. d. thesis, Universidad Politécnica de Cataluña (2009)
5. Mason, M.T.: Mechanics of robotic manipulation. The MIT Press, Cambridge (2001)
6. Kao, I., Lynch, K., Burdick, J.: Contact modeling and manipulation, pp. 647–668. Springer, Berlin (Sep 2008). Springer Handbook of Robotics
7. Roa, M., Suarez, R.: Computation of independent contact regions for grasping 3-d objects. IEEE Trans. Robot. **25**(4), 839–850 (Aug 2009)
8. Salisbury, J.: Kinematics and force analysis of articulated hands. Ph.d. thesis, Standford University (1982)

9. Mason, M.T., Salisbury, J.K.: Robot hands and the mechanics of manipulation. The MIT Press, Cambridge (1985). The MIT series in Artificial Intelligence
10. Cutkosky, M.: On grasp choice, grasp models, and the design of hands for manufacturing tasks. IEEE Trans. Robot. Autom. **5**(3), 269–279 (Jun 1989)
11. Lin, Q., Burdick, J.M., Rimon, E.: A stiffness-based quality measure for compliant grasps and fixture. IEEE Trans. Robot. Autom. **16**(6), 675–688 (2000)
12. Bicchi, A.: On the problem of decomposing grasp and manipulation forces in multiple whole-limb manipulation. Int. J. Robot. Auton. Syst. **13**, 127–147 (1994)
13. Harada, K., Kaneko, M., Tsuji, T.: Rolling based manipulation for multiple objects. In: Proceedings. ICRA '00. IEEE International Conference on Robotics and Automation. vol. 4, pp. 3887–3894 (2000).
14. Olsson, H., Astrom, K., de Wit, C., Gafvert, M., Lischinsky, P.: Friction models and friction compensation. Eur. J. Control **4**(3), 176–195 (1998)
15. Canudas de Wit, C., Olsson, H., Astrom, K., Lischinsky, P.: A new model for control of systems with friction. IEEE Trans. Autom. Control **40**(3), 419–425 (Mar 1995)
16. Erdmann, M.: On a representation of friction in configuration space. Int. J. Robot. Res. **13**(3), 2179–2184 (1994)
17. Lynch, K.: Toppling manipulation. In: Proceedings. 1999 IEEE International conference on robotics and automation, 1999, vol. 4, pp. 2551–2557 (1999)
18. Tao Zhang, M., Goldberg, K., Smith, G., Beretty, R.P., Overmars, M.: Pin design for part feeding. Robotica **19**(06), 695–702 (Sep 2001)
19. Maeda, Y., Arai, T.: Planning of graspless manipulation by a multifingered robot hand. Adv. Robot. **19**(5), 501–521 (2005)
20. Ciocarlie, M., Miller, A., Allen, P.: Grasp analysis using deformable fingers. In: IEEE/RSJ International Conference on Intelligent Robots and Systems, In (2005). (IROS 2005). pp. 4122–4128 (Aug 2005)
21. Ciocarlie, M., Lackner, C., Allen, P.: Soft finger model with adaptive contact geometry for grasping and manipulation tasks. In: EuroHaptics Conference and Symposium on Haptic Interfaces for Virtual Environment and Teleoperator Systems. Second joint, pp. 219–224 (Mar 2007)
22. Howe, R., Kao, I., Cutkosky, M.: The sliding of robot fingers under combined torsion and shear loading. In: Proceedings IEEE International Conference on Robotics and Automation, pp. 103–105 vol. 1 (Apr 1988)
23. Kao, I., Cutkosky, M.R.: Quasistatic manipulation with compliance and sliding. Int. J. Robot. Res. **11**(1), 20–40 (Feb 1992)
24. Howe, R.D., Cutkosky, M.R.: Practical force-motion models for sliding manipulation. I. J. Robot. Res. **15**(6), 557–572 (1996),http://dblp.unitrier.de/db/journals/ijrr/ijrr15.html#HoweC96
25. Kao, I., Yang, F.: Stiffness and contact mechanics for soft fingers in grasping and manipulation. IEEE Trans. Robot. Autom. **20**(1), 132–135 (Feb 2004)
26. Pérez-González, A., Fenollosa-Esteve, C., Sancho-Bru, J.L., Sánchez-MaríÂnn, F.T., Vergara, M., RodríÂnguez-Cervantes, P.J.: A modified elastic foundation contact model for application in 3d models of the prosthetic knee. Med. Eng. Phy. **30**(3), 387–398 (2008), http://www.sciencedirect.com/science/article/pii/S1350453307000616
27. Hertz, H.: On the contact of rigid elastic solids and on hardness. Ch 6: Assorted Papers (1882)
28. Johnson, K.L.: Contact mechanics. Cambridge University Press, Cambridge (1995)
29. Xydas, N., Kao, I.: Modeling of contact mechanics and friction limit surfaces for soft fingers in robotics, with experimental results. Int. J. Robot. Res. **18**(9), 941–950 (1999)
30. Xydas, N., Bhagavat, M., Kao, I.: Study of soft-finger contact mechanics using finite elements analysis and experiments. In: Proceedings 2000 ICRA. Millennium Conference. IEEE International Conference on Robotics and Automation. Symposia Proceedings (Cat. No. 00CH37065) (April), pp. 2179–2184 (2000)
31. Gonthier, Y.: Contact dynamics modelling for robotic task simulation. Ph.D. thesis, University of Waterloo (Oct 2007)

32. Shimoga, K.B.: Robot grasp synthesis algorithms: A survey. I. J. Robot. Res. **15**(3), 230–266 (1996), http://dblp.uni-trier.de/db/journals/ijrr/ijrr15.html#Shimoga96
33. Bicchi, A.: On the closure properties of robotic grasping. Int. J. Robot. Res. **14**, 319–334 (1995)
34. Nguyen, V.D.: Constructing force-closure grasps. Institute of Electrical and Electronics Engineers (1988)
35. Salisbury, J.K., Roth, B.: Kinematic and force analysis of articulated mechanical hands. J. Mech. Trans. Autom. Des. **105**(1), 35–41 (1983), http://link.aip.org/link/?JMT/105/35/1
36. Mishra, B., Schwartz, J.T., Sharir, M.: On the existence and synthesis of multifinger positive grips. Algorithmica **2**(1–4), 541–558 (Nov 1987)
37. Hirai, S., Asada, H.: Kinematics and statics of manipulation using the theory of polyhedral convex cones. Int. J. Robot. Res. **12**(5), 434–447 (October 1993)
38. Xiong, Y.L.: Theory of point contact restraint and qualitative analysis of robot grasping. Sci. China (Scientia Sinica) Series A **37**(5), 629–640 (1994), http://www.scopus.com/inward/record.url?eid=2-s2.0-0001723889&partnerID=40&md5=39ad9b3ac9ce43daa95fced360cb6d18
39. Ferrari, C., Canny, J.: Planning optimal grasps. In: Proceedings 1992 IEEE International Conference on Robotics and Automation, pp. 2290–2295 (1992)
40. Zhu, X., Wang, J.: Synthesis of force-closure grasps on 3-d objects based on the q distance. IEEE Trans. Robot. **19**(4), 669–679 (2003), http://dblp.uni-trier.de/db/journals/trob/trob19.html#ZhuW03
41. Nakamura, Y., Nagai, K., Yoshikawa, T.: Dynamics and stability in coordination of multiple robotic mechanisms. Int. J. Robot. Res. **8**(2), 44–61 (1989), http://ijr.sagepub.com/content/8/2/44.abstract
42. Trinkle, J.: On the stability and instantaneous velocity of grasped frictionless objects. IEEE Trans. Robot. Autom. **8**(5), 560–572 (Oct 1992), http://ieeexplore.ieee.org/xpl/articleDetails.jsp?tp=&arnumber=163781&contentType=Journals+%26+Magazines&searchField%3DSearch_All%26queryText%3Don+the+stability+and+instantaneous+velocity+of+grasped
43. Liu, Y.H.: Qualitative test and force optimization of 3-d frictional form-closure grasps using linear programming. IEEE Trans. Robot. Autom. **15**(1), 163–173 (Feb 1999), http://ieeexplore.ieee.org/xpl/articleDetails.jsp?tp=&arnumber=744611&contentType=Journals+%26+Magazines&searchField%3DSearch_All%26queryText%3DQualitative+test+and+force+optimization+of+3-D+frictional+formclosure+grasps+using+linear+programming
44. Han, L., Trinkle, J.C., Li, Z.X.: Grasp analysis as linear matrix inequality problems. IEEE Trans. Robot. **16**(6), 663–674 (2000), http://dblp.uni-trier.de/db/journals/trob/trob16.html#HanTL00
45. Zhu, X., Ding, H., Wang, M.: A numerical test for the closure properties of 3-d grasps. IEEE Trans. Robot. Autom. **20**(3), 543–549 (June 2004)
46. Zheng, Y., Chew, C.M.: Distance between a point and a convex cone in —dimensional space: computation and applications. IEEE Trans. Robot. **25**(6), 1397–1412 (Dec 2009)
47. Zheng, Y., Lin, M., Manocha, D.: On computing reliable optimal grasping forces. IEEE Trans. Robot. **28**(3), 619–633 (June 2012)
48. Buss, M., Hashimoto, H., Moore, J.: Dextrous hand grasping force optimization. IEEE Trans. Robot. Autom. **12**(3), 406–418 (June 1996), http://ieeexplore.ieee.org/xpl/articleDetails.jsp?tp=&arnumber=499823&contentType=Journals+%26+Magazines&sortType%3Dasc_p_Sequence%26filter%3DAND(p_IS_Number%3A10865)
49. Buss, M., Faybusovich, L., Moore, J.B.: Dikin-type algorithms for dextrous grasping force optimization. Int. J. Robot. Res. 17(8), 831–839 (1998), http://ijr.sagepub.com/content/17/8/831.abstract
50. Helmke, U., Huper, K., Moore, J.: Quadratically convergent algorithms for optimal dextrous hand grasping. IEEE Trans. Robot. Autom. **18**(2), 138–146 (Apr 2002)
51. Liu, G., Li, Z.: Real-time grasping-force optimization for multifingered manipulation: theory and experiments. IEEE/ASME Trans. Mechatron. **9**(1), 65–77 (Mar 2004)
52. Gazeau, J.P., Zeghloul, S., Ramirez, G.: Manipulation with a polyarticulated mechanical hand: a new efficient real-time method for computing fingertip forces for a global manipulation strategy. Robotica **23**, 479–490 (6 2005), http://dx.doi.org/10.1017/S0263574704001067

53. Cornellà, J., Suárez, R., Carloni, R., Melchiorri, C.: Dual programming based approach for optimal grasping force distribution. Mechatronics **18**(7), 348–356 (2008), http://www.sciencedirect.com/science/article/pii/S0957415807001080
54. Zheng, Y., Qian, W.H.: Limiting and minimizing the contact forces in multifingered grasping. Mech. Mach. Theory **41**(10), 1243–1257 (2006), http://www.sciencedirect.com/science/article/pii/S0094114X05001953
55. Morales, A., Sanz, P., Del Poblil, A., Fagg, A.: Vision-based three-finger grasp synthesis constrained by hand geometry. Robot. Auton. Sys. **54**(6), 496–512 (2006)
56. Ponce, J., Faverjon, B.: On computing three-finger force-closure grasps of polygonal objects. IEEE Trans. Robot. Autom. **11**(6), 868–881 (Dec 1995)
57. Ponce, J., Sullivan, S., Sudsang, A., Boissonnat, J.D., Merlet, J.P.: On computing four-finger equilibrium and force-closure grasps of polyhedral objects. Int. J. Robot. Res. **16**(1), 11–35 (1997)
58. Liu, Y.H.: Computing n-finger form-closure grasps on polygonal objects. Int. J. Robot. Res. **19**(2), 149–158 (2000), http://ijr.sagepub.com/content/19/2/149.abstract
59. Napier, J.: The prehensile movements of the human hand. Surger **38**(4), 902–913 (1956)
60. Stansfield, S.: Robotic grasping of unknown objects: a knowledge-based approach. Int. J. Robot. Res. **10**(4), 314–326 (1991)
61. Wren, D., Fisher, R.: Dextrous hand grasping strategies using preshapes and digit trajectories. In: IEEE International Conference on Systems, Man and, Cybernetics. vol. 1, pp. 910–915 vol. 1 (Oct 1995)
62. Miller, A.T., Knoop, S., Christensen, H., Allen, P.K.: Automatic grasp planning using shape primitives. In: Proceedings ICRA'03. IEEE International Conference on Robotics and Automation, vol. 2, pp. 1824–1829. IEEE (2003)
63. Miller, A., Allen, P.: Graspit! a versatile simulator for robotic grasping. IEEE Robot. Autom. Mag. **11**(4), 110–122 (Dec 2004)
64. Diankov, R.: Automated construction of robotic manipulation programs. Ph.D. thesis, Carnegie Mellon University, Robotics Institute (Aug 2010)

Chapter 3
Robot Grasping Simulation

3.1 Introduction

Robot simulators have accompanied robotics for a long time and have been an essential tool for the design and programming of industrial robots. Almost all industrial manipulator manufacturers offer simulation packages accompanying their robotic products. These tools allow the users to program and test their applications without using the real hardware or even building it since such tools allow the analysis of behaviours and performance beforehand.

In this work, simulation is understood as a tool which will enable us to study and improve robotic grasping. The simulator would be capable, not only of replacing the real hardware, but, more importantly, of being part of the robot's cognitive system, acting as a prediction engine able to emulate the retrieval of self knowledge.

The role of simulation as a prediction engine has been studied in psychology in the field of *grounded cognition*. There, simulation is devised as "the re-enactment of the perceptual, motor, and introspective states acquired during experience with the world, body, and mind" [1]. Studies in this area suggest that simulation provides a core form of computation in the human brain with a diverse collection of simulation mechanisms that supports the brain cognitive activities, including high-level perception, implicit memory, working memory, long-term memory, and conceptual knowledge.

Based on these approaches, the simulator is used as a memory and reasoning agent in the robot applications of this work. It will rests upon the Predict-Act-Perceive paradigm in which two loops run in parallel (not necessarily synchronously): One in the real world and one in simulation (see Fig. 3.1).

The current state of the world is first observed with different real sensors and used to build the world model in the simulator. The simulator then becomes the internal world of the robot in which we can have two different modes of prediction:

- Predict the values of different sensors, given an exploratory action and potential assumptions.

B. León et al., *From Robot to Human Grasping Simulation*,
Cognitive Systems Monographs 19, DOI: 10.1007/978-3-319-01833-1_3,

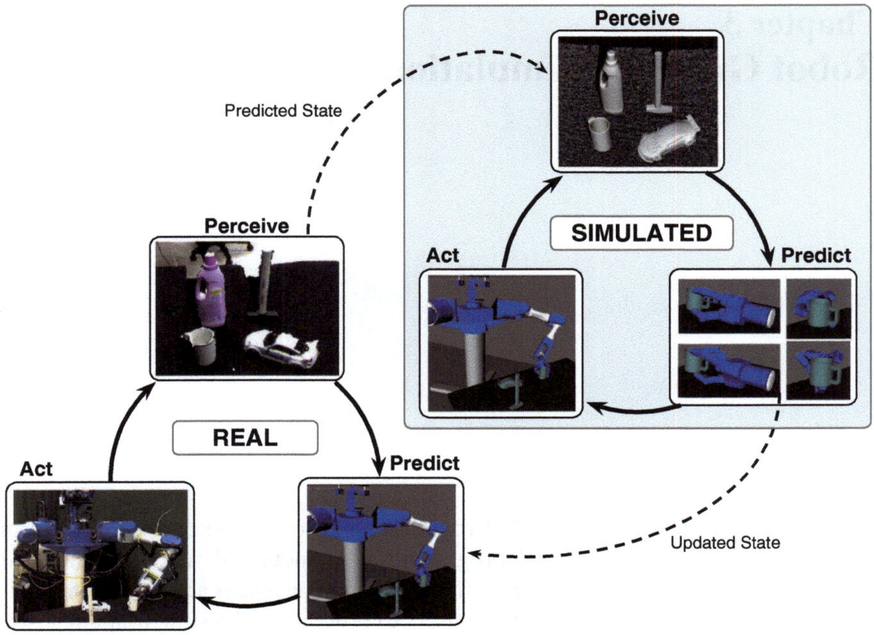

Fig. 3.1 Predict-Act-Perceive paradigm

- Predict the outcome of a specific action applied to the world. When interacting with objects, simulation can provide the estimation of the next state, given measurements of tactile sensors after a grasp has been applied to an object and predict the outcome of a lifting action. When interacting with objects, simulation can provide the estimation of the next state that can ease, for example, tracking of objects.

Once the prediction has been made by the simulator, the selected action can be applied in reality and the real sensor values can be compared with the predicted ones. A mismatch can be treated as a surprise that may, for example, trigger an exploratory action.

Thus, the paradigm encompasses on the one hand the evaluation of a currently executed action based on its predicted outcome in simulation, while on the other hand the simulation state is grounded and refined based on the experience in the real world, which consequently leads to an improvement in the prediction and parameterization of future actions.

3.1.1 Requirements for a Grasp Simulator

In order to be able to act as a *grasp reasoning engine*, the simulator needs to fulfil multiple requirements from the robotic community. From a scientific point of view, a novel simulator for robot grasping should provide primarily a realistic simulation of dynamic properties of, at least, rigid objects and advanced contact models including soft contacts and deformable surfaces. From a practical and user point of view, it should include the models of the most popular robot hands, and provide the possibility of easily creating and adding new ones. Furthermore, it should provide realistic simulations of real actuators and sensors, which would enable the use of the same API as in their real counterparts. Regarding sensors, a grasping simulator has to provide simulations of specific grasping sensors, such as force/torque, contact and tactile. Finally, it should provide a rich and detailed visualization of simulations.

With respect to software engineering, a novel robot grasping simulator must be implemented in a modular way that enables on the one hand, an easy extension by both developers and users and on the other hand, the integration with commonly-used robotics software frameworks. In order to increase its opportunities of being adopted and used in the scientific community, the simulator should be open source and make use of open standards for file formats and other representations. In addition, the simulator should have appropriate tools to import/export robot and object models to/from standard representations.

The problems involved in the general requirements allow the definition of the following particular requirements:

- **Modularity and extendibility**: A modular architecture that clearly separates the different functionalities of the system. All these modules should be coordinated through a common representation of the parts in the environment, also allowing inclusion of new modules or plug-ins. It should allow the replacement of actual hardware components by software emulations.
- **Robot hand modelling**: Simulations of popular mechanical devices and environment parts should be adapted for the simulator. This include the most popular dexterous hands and a model of the human hand.
- **Sensor modelling**: An important part of the algorithms for dexterous manipulation considers the use of sensors. The simulator should include the design of virtual sensors that emulate the information provided by the real sensors. Types of sensors that are important for grasping include arrays of pressure cells and 6D force/torque sensors.
- **Contact and friction models**: A key feature of a manipulation simulator is the modelling of the interactions between the objects. This interaction is expressed in terms of the forces and torques transmitted through these contacts. The simulator should consider different types of contacts (punctual, edge, and surface), and different models of friction (frictionless, coulomb). In a second stage soft compliant contact models should be considered.
- **Static and dynamic modelling**: The static model of an interaction considers the contacts, forces and moments in a set of objects (robots included) and produces a snapshot of the state. In the dynamic model, the velocities, and accelerations produced by the applied forces are also considered; masses and inertias are needed too. The simulator should include modular engines that compute the static and dynamic properties of the simulated systems.

- **Import/export of models from/to standard representations**: The ability to easily import and export models from and to other software tools.
- **Open source development**: A development procedure based on a publicly available code that help to the involvement of other groups. This collaboration should consist in the software testing, bug reporting, solution proposal, and even patch development, leading to a product that best fits the necessities of users.
- **Visualise current knowledge about the world**: It should be able to stand as a stand-alone program for simulations of possible actions of the current robot hardware and their interaction with the world objects.
- **Make predictions**: It should apply its a priori knowledge about the kinematics embedded in its physics engine to the world representation (foreground/ background) in order to generate a realistic simulation of operation in a given environment. An extension of this role should be the use of the engine for task planning.

3.1.2 Related Work

In order to develop a tool that meets the previous requirements, a basic practical principle was adopted: *Do not reinvent the wheel.* This means, first to review the existing software paying special attention to those that already meet part of the requirements and, second to make use of existing open and widely-available software packages and standards.

Often, simulation tools used to support research are specifically developed for particular experiments. However, there have been some successful attempts to develop general robot simulators specifically for mobile robotics.

Stage and Gazebo are respectively 2D and 3D simulator back-ends for Player [2], which is a widely used free software robot interface. Gazebo [3] in particular, implements a 3D multi-robot simulator which includes dynamics for outdoor environments. It implements several robot models, actuators and sensors. USARSim [4] has a similar functionality. It is a free mobile robot simulator based on a gaming engine. Microsoft provides its Robotics Studio [5], a framework for robot programming that includes a visual simulation environment. OpenHRP [6] is an open software platform with various modules for humanoid robot systems such as a dynamics simulator, a view simulator, motion controllers and motion planners. OpenHRP is integrated with CORBA, with each module, including the dynamics simulator implemented as a CORBA server. Commercial options include Webots [7], a product which has been widely successful in educational settings.

The variety of simulation tools for robotic grasping is rather limited. The most renowned and prominent one is *GraspIt!*, a grasping simulation environment [8]. *GraspIt!* includes a general framework to model robot hands and arms, especially joint modelling, a collision solver, a friction hard contact modelling, an engine to emulate rigid objects dynamics, and a trajectory controller. However it fails to provide soft contact models, and the emulation of the most common kinds of sensors such as pressure and force/torque. Its monolithic architecture makes its extension, improvement and integration with other tools very difficult. In addition, it does not provide a convenient Application Programming Interface (API), which allows script programming.

Recently, [9] have developed a MATLAB toolbox for the analysis of grasping objects both for robotic and human hands. It focuses on using synergies to model the coupling between joints induced by underactuated hand control. However, it is a new toolbox and it is still under development.

Another existing and publicly available software framework is OpenRAVE, the Open Robotics and Animation Virtual Environment [10]. It has been designed as an open architecture targeting a simple integration of simulation, visualization, planning, scripting and control of robot systems. It has a modular design, which allows its extension and further development by other users. Regarding robot grasping simulation, it provides similar functionality to *GraspIt!* and various path planning components. It provides the models of several robot arms and hands and allows the integration of new ones. It also enables the development of virtual controllers for such models.

The quality of a simulation is largely dependent on the *physics engine* which calculates the dynamics of the simulation, and the *rendering engine* which is used to visualize it. The results of the physics simulation are highly dependent on the accuracy of the models which are provided by the user. There are many physics engines available with varying quality and cost. These packages usually include collision checkers, friction and contact models and a varied type of motion constraints to enable the simulation of articulated bodies. Their flexibility makes them a suitable tool to simulate the dynamics behaviour of a robot, without the need to reformulate and implement the underlying physics models. ODE (Open Dynamics Engine) is the most popular rigid body dynamics implementation for robotics simulation applications being used in most of the previously mentioned simulators. Other free open source engines are Bullet, Newton Game Dynamic and dvc3D [11]. However, they have several limitations which limit their ability to accurately reproduce dynamics simulation [12].

Similarly, a wide variety of 3D rendering engines also exist. The game industry has helped to advance the quality of these engines to its current limits; to the point where open-source engines that provide this exceptionally high-quality visualization are now available.

To our knowledge, none of the existing simulation tools and software packages fulfil all of the above-mentioned requirements. In order to achieve them, we have developed a software toolkit specifically designed for grasping simulation, called OpenGRASP. However, we adopted the tools that already meet part of the requirements and, that were open and widely-available software packages and standards. After the review of existing simulators, we concluded that OpenRAVE is the tool that most closely meets the requirements. Therefore the efforts were focused on improving and extending OpenRAVE capabilities and features towards the realization of an advanced grasping simulator. In the following sections, a brief introduction to OpenRAVE is presented.

3.2 OpenRAVE

The core of the toolkit is OpenRAVE[1] [10], a planning architecture designed for autonomous robot applications. It consists of three layers: a core, a plugins layer for interfacing to other libraries, and scripting interfaces for easier access to functions (Fig. 3.2).

3.2.1 The Core Layer

The main API is coded in C++ using the Boost C++ libraries[2] as a solid basis of low-level management and storage structures. The Boost flavours of shared pointers allow object pointers to be safely reference counted in a heavily multi-threaded environment.

3.2.2 The Scripting Layer

It enables network scripting environments like Octave, Matlab and Python to communicate with the core layer in order to control the robot and the environment. It is possible to send commands to change any aspect of the environment, read any of its information, move real robots, or change physics/collision libraries. The scripts also enable the control of multiple OpenRAVE instances across the network, thus allowing different users to independently see and interact with the environment.

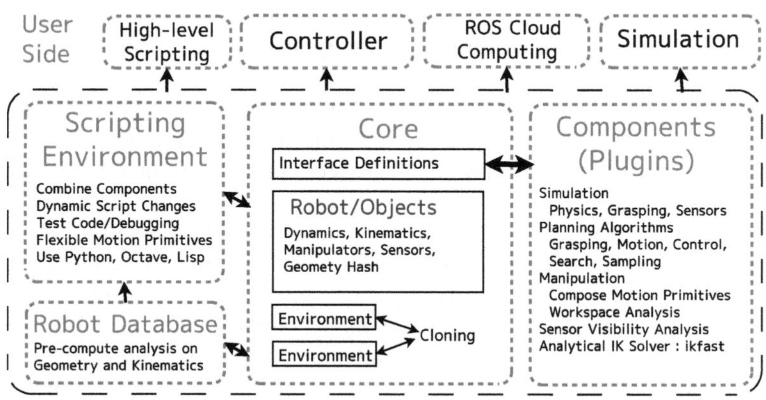

Fig. 3.2 OpenRAVE architecture reproduced from [10]

[1] http://www.openrave.org/docs/latest_stable/

[2] http://www.boost.org/

3.2.3 *The Plugins Layer*

OpenRAVE is designed as a plugin-based architecture, which allows the creation of new components to continuously improve its original specifications. Each plugin is an implementation of a standard interface that can be loaded dynamically without the need to recompile the core. Following this design, different plugins can be created for components such as sensors, planners, controllers or physics engines. The core layer communicates with the hardware through the plugins using more appropriate robotics packages such as Player and the Robot Operating System (ROS).[3]

A GUI can be optionally attached to provide a 3D visualization of the environment. It periodically queries the core to update the world's view and allows the user to change the position of the objects in the scene. As viewers are provided through plugins, a single OpenRAVE instance can allow multiple viewers to communicate with multiple copies of the environment.

3.2.4 *Object Manipulation in OpenRAVE*

Although OpenRAVE can be used for different robot applications, its main focus is to provide an architecture for manipulation planning purposes. As this is also the focus of this book, an introduction of how OpenRAVE deal with manipulation tasks is presented which also lays down the basic terminology for the following sections.

A manipulation task is defined in OpenRAVE as moving a single object to achieve a set of goals. It can be either to place an object in a designated location or to act on the an object by moving it a relative distance from its initial configuration.

In order to perform such tasks, several information about the environment, the manipulable objects and the robot should be specified. Figure 3.3 presents the parts of a scene that OpenRAVE needs to define a manipulation task. A brief description of the required information the user should specify to the system for each task is presented in Table 3.1.

3.3 OpenGRASP: Simulation Toolkit

In this section, the different components of *OpenGRASP* are presented. The core of the toolkit is an improved version of OpenRAVE with different plugins developed to add new functionality. It has also been enhanced with a Robot Editor and the adoption of the COLLADA file format. In the last part of the chapter, tactile sensor simulation is studied in detail, as a new tactile sensor model—of crucial importance for grasping—was developed which utilizes collision detection and response methods using soft contacts as well as a full friction description.

---

[3] http://www.ros.org

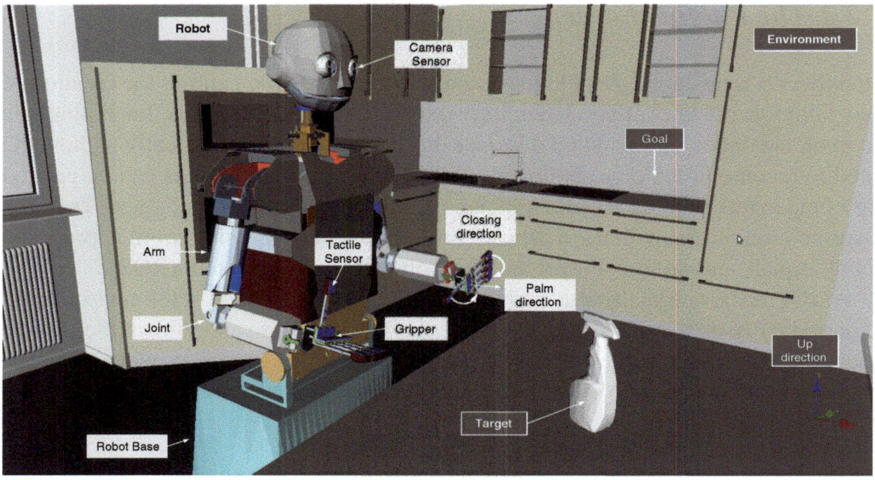

Fig. 3.3 Parts of an OpenRAVE scene

Table 3.1 OpenRAVE specification parameters

	Parameter	Definition
Environment		Geometry, the up direction, and the regions of possible robot movement
Goals		Goal positions to move the target/robot to
Target	Geometry	A set of rigid links (using convex decompositions) connected by joints
	Constraints	A function defining the valid configurations of the target
Robot	Geometry	A set of rigid links defining the robot shape
	Kinematics	Joints connecting links that define the configuration space and its limits
	Dynamics	The dynamic properties of all the links and joints, parameter limits
	Arms	A chain of joints defining the arm and the type of invese kinematics to use
	Grippers	A set of joints controlling the fingers, joint directions for closing fingers
	Manipulators	Each manipulator consist of an arm and a gripper
	Sensors	Type of sensor, robot link and location attached to
	Robot Base	How robot base moves across the environment. Moves all manipulators and sensors
	Control	Expected positioning errors on physical movement from inputs, and hardware limits

A diagram of the OpenGRASP architecture is presented in Fig. 3.4, highlighting in blue the contributions made to existing software. These extensions are described in more detail in the following sections.

3.3.1 Developed Plugins

Although many plugins were already implemented in OpenRAVE to provide basic functionality, the grasp simulation functionality had several shortcomings. In order to make OpenRAVE suitable for grasping simulation, it was required:

- Implementation of plugins for specific sensors used to improve the grasping capabilities of the robot.
- Implementation of more physics engines and collision checkers that help to compare and improve simulation performance.
- Implementation of a standard plugin interface for a basic actuator and implementations of different motors. This would allow us to accurately simulate the robot's arm and hand articulations.

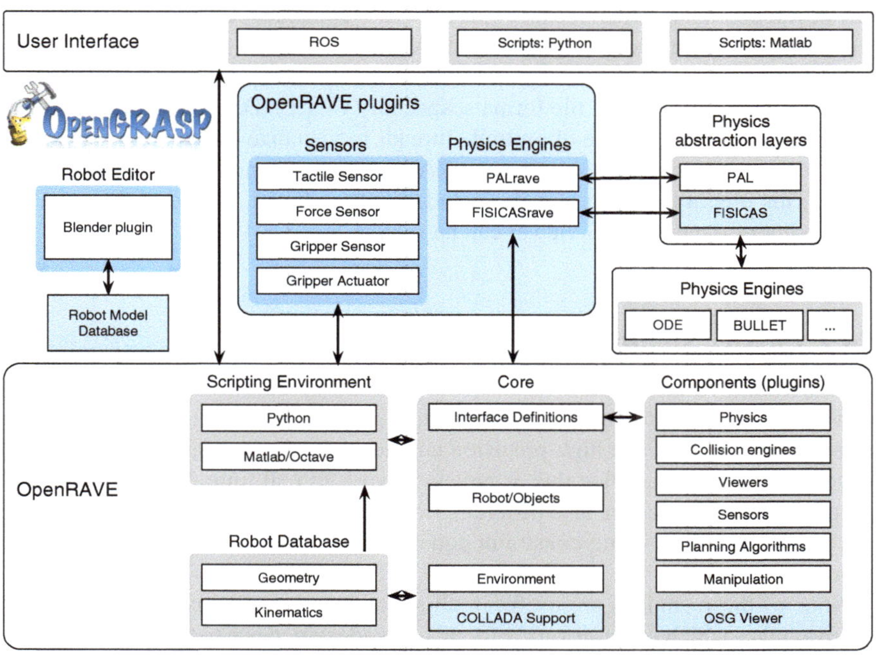

Fig. 3.4 System architecture: components part of the OpenGRASP toolkit (*blue*) and their interface with other available software packages (*gray*)

These considerations were taken into account in the toolkit design. Two new sensor plugins were developed to be used mainly for anthropomorphic robot hands. One is a tactile sensor, commonly used in finger tips such as in the Barrett hand, which detects and calculates the forces on the predefined sensory area and returns them as an array (see Sect. 3.4). The other is a force sensor, placed for example in the wrist, to measure the forces applied while grasping. The physics engine was queried for the force calculated on the sensor's body. Additionally, as models for actuators were not included in OpenRAVE, a new sensor type called ST_Actuator was developed.

In order to simulate robot grippers, a *Gripper Sensor* plugin was developed which returns the distance between fingers, their current and velocity. Additionally, a model of the *Gripper Actuator* was also included which can control the velocity of the fingers and the maximum current applied.

In order to use different physics engines, new plugins were implemented which make use of the physics abstraction layer PAL [13] and FISICAS, which are addressed in more detail in the following section.

Visualisation is an important part of the simulation. OpenRAVE uses by default Coin3D/Qt[4] to render the environment, which has several disadvantages. It has recently changed its license to GPL which poses a lot of limitations on its distribution and it doesn't support multi-thread viewers. Therefore, a new viewer plugin was developed using OpenSceneGraph (OSG).[5] OpenSceneGraph [14] is an open source high performance 3D graphics toolkit, written in C++ and OpenGL which can be used in the most common platforms such as Windows, OSX and GNU/Linux. It enables the use of several file formats, specially COLLADA, VRML and IV files. Moreover, it enables the use of multiple threads to visualize different environments simulating concurrent processes in OpenRAVE, like the planning of different trajectories. This plugin has been already included in the official release of OpenRAVE. A snapshot of the new OSG viewer can be seen in Fig. 3.5.

3.3.2 Physics Simulation

Nowadays, there exist many available physics engines, both commercial and open-source. Some of them are high-precision engines that require higher computational power while others sacrifice this accuracy to work in real time. The methods they use to simulate physics are also different. Some of them use penalty methods, some rely on physical laws using constraint equations, and others use methods based on impulses.

None of these engines are useful in all situations, they all have advantages and disadvantages which makes it very difficult to decide which one to use for a simulator.

[4] http://www.coin3d.org/

[5] http://www.openscenegraph.org/

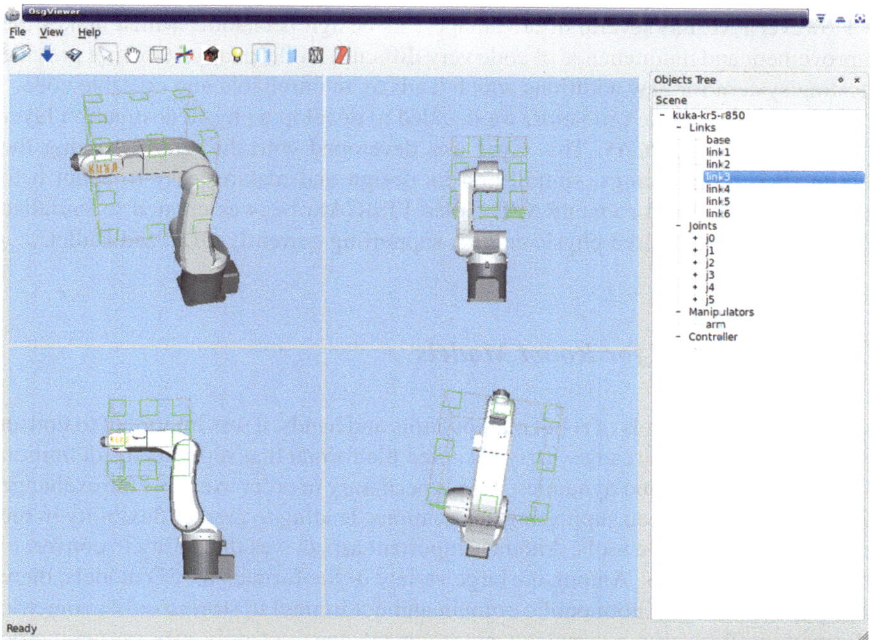

Fig. 3.5 Screenshot of OpenRAVE using the new osgviewer plugin

It basically depends on what we want to simulate in addition to what the application of the simulator will be.

In order to simplify this decision, a physics abstraction layer appears as a solution. It provides an interface to a number of different physics engines allowing us to dynamically switch between them. This functionality adds flexibility to the simulator offering the possibility to, depending on the specific environment and use, decide which engine would provide the best performance. It is also possible to test and compare the behaviour of different engines.

The OpenRAVE Physics Engine interface allows the simulator to run using different engines. It also has an interface to implement different collision checkers. Each one of them has to be created as a plugin, extending either the PhysicsEngineBase or the CollisionCheckerBase class.

The Physics Abstraction Layer (PAL)[6] [13] is a software package created by Adrian Boeing that saves us from having to decide at the start what physics engine to use. A plugin to use PAL within OpenRAVE was created, called palrave. This plugin is able to initialize PAL with the specific engine we want to use, without the need of creating different plugins for each one of them.

[6] http://pal.sourceforge.net/

However PAL has several disadvantages. Its design is chaotic, which makes the improvement and maintenance of code very difficult. Additionally, it did not provide a testing system for new additions which leads to an unreliable and unstable code.

After realising these problems, we decided to develop a physic abstraction layer which we called *FISICAS*. This layer was developed with the aim of solving the current PAL shortcomings, simplifying its design and making easy to adapt it to our needs. A plugin for OpenRAVE called FISICASrave was created to initialize FISICAS with the desired physic engine, supporting currently ODE and Bullet.

3.3.3 File Format for Robot Models

For the storage of models of robot manipulators and hands, it was important to find an open, extensible and already widely accepted file format that supports the definition of at least kinematics and dynamics. This is necessary in order to enable the exchange of robot models between supporting applications, leading to greater flexibility in the selection of appropriate tools. Another important aspect was the ability to convert to and from other formats. Among the large variety of file formats for 3D models, there are only a few that are both public domain and not limited to storing only geometric information. Typically, a simulator environment does not only rely on geometric structures but also, for instance, on information about dynamics, kinematics, sensors and actuators of the robot.

COLLADA,[7] a format widely distributed and accepted as an industry standard (3D Studio, Blender, OSG, OGRE, Sony, etc.) was chosen as the preferred file format for the simulator. It has a clear and extensible design, and there are open source frameworks available that facilitate its integration into new applications.

Since version 1.5, the standard contains many useful constructs dedicated to describing kinematic chains and dynamics that can be used directly for the description of robot models. COLLADA is an XML-based file format that enables and encourages developers to extend the specification to their needs without having to violate the underlying schema definition.

In order to support specific robot features like sensors and actuators, we have used this mechanism to extend COLLADA partially using the original OpenRAVE file definition. These additions are specific to the simulator and are hidden to all other applications so that compatibility remains guaranteed. Support for COLLADA import and export has been included in the official release of OpenRAVE.

[7] https://collada.org/

3.3.4 Robot Editor

With the creation of a simulator for grasping the need also arises for a large data base of geometrical, kinematic and dynamic models of robot arms and manipulators. To fill this gap, a modelling tool called the *Robot Editor* was developed by Stefan Ulbrich as part of OpenGRASP. Its main goal is to facilitate modelling and integration of many popular robots. The development is driven by the following key aspects:

- **Geometric modelling**: The creation of new robots models requires a tool that excels in modeling of the geometric components (i.e., meshes).
- **Semantic modelling**: Even more important is the ability to allow the description of semantic properties, such as definitions of kinematic chains, sensors and actuators, or even specify algorithms.
- **Dynamics modelling**: Another necessary aspect is the ability to define physical attributes of the robot's elements. At the moment, the focus lies on the dynamics of rigid bodies.
- **Conversion**: Robot models usually come in a variety of different file formats. The modelling tool needs to be capable of processing these formats and converting them into the COLLADA standard. GraspIt! files in particular, being an already widely-used standard with many conform models available, should be readily usable by the simulator.

The conceptual design of the Robot Editor relies on two techniques: on the one hand the open data format COLLADA and on the other hand on the open source project Blender. Blender is a very versatile, powerful and extensible 3D editor that has been chosen because of its convenient 3D modelling capabilities and the built-in support for many CAD file formats and rigid body kinematics. Furthermore, it can be easily extended via a Python scripting interface and offers high-quality ray tracing.

Blender itself, however, lacks the functionality and the appropriate interface for the convenient definition of robot components. In addition, conversions between certain file formats need to be improved or implemented, namely the import of the COLLADA format and GraspIt! robot models.

The scripting mechanism allows the creation of user-interface that is highly specialized for use in robotics (see Fig. 3.6a). COLLADA support in Blender was limited to documents in the older specification 1.4 which excludes the newly introduced kinematics and dynamics. The additional data required by the simulator also needs to be included in the resulting document. This led to the further development of COLLADA compatibility which now enables the Robot Editor to create valid documents suitable for simulation. Figure 3.6a shows a functional model of the Karlsruhe anthropomorphic hand [15] modified using the Robot Editor and Fig. 3.6b the resulting COLLADA file loaded into the simulator.

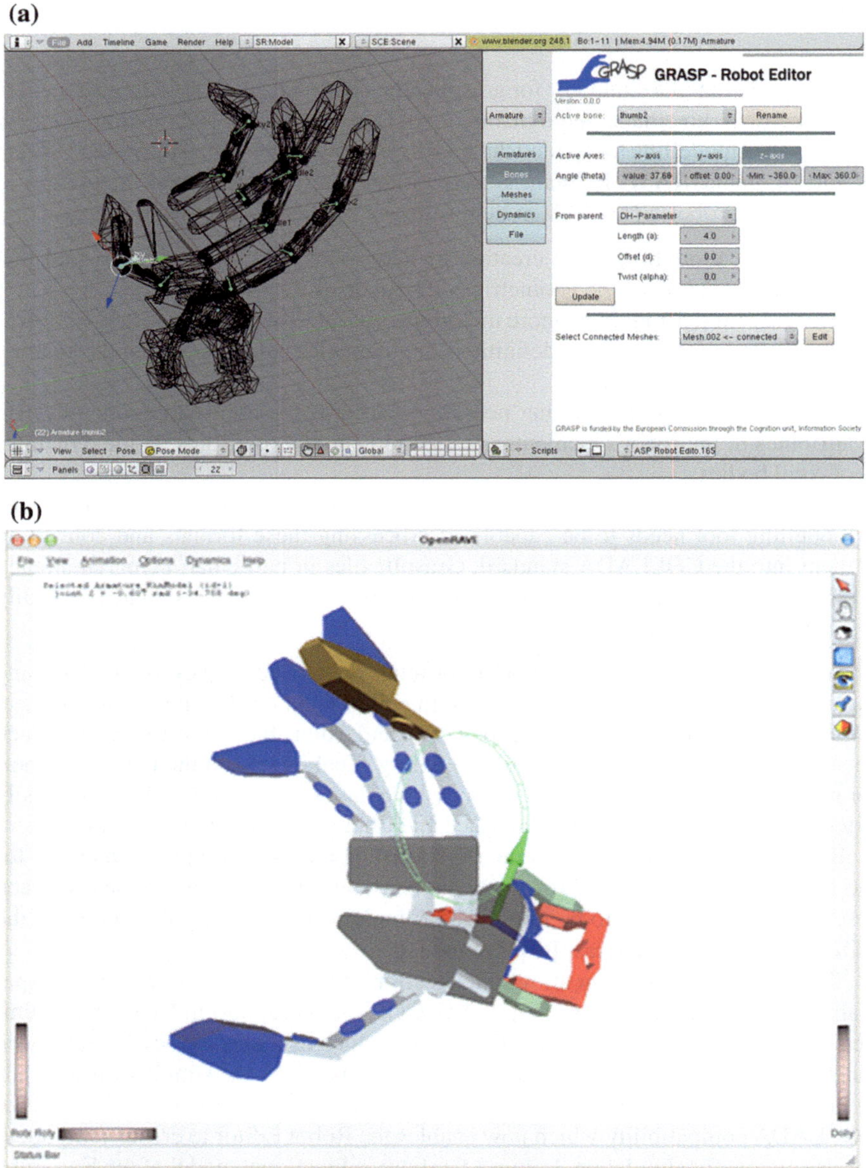

Fig. 3.6 Modeling the Karlsruhe anthropomorphic robot hand. **a** The user interface of the Robot Editor and **b** Screenshot of the complete model in the simulator

3.3.5 Robot Models

As stated in the simulator requirements, it is of great importance to have models of the most popular robot hands included in the toolkit. The modeling capabilities of the Robot Editor already enable quick and easy creation of new models. Therefore the models of the most popular robotic hands were created to be available for OpenRAVE as part of the OpenGRASP toolkit. Additionally, several complete robot platforms available at different European labs were also modelled.

3.3.5.1 Models of Popular Robot Hands

So far, a selection of robot hands has been transformed into COLLADA 1.5 for use within the simulator (see Fig. 3.7). In addition to these new models, there are various models available in the older XML file format which is still supported.

3.3.5.2 Models of Complete Robot Platforms

Additionally, we simulated the models of the robot platforms available within the laboratories of the partner members of the GRASP project. These platforms are used in the applications described in the following chapters of this book.

Schunk PG70 Gripper

We have also modeled the platform at the Lappeenranta University of Technology (LUT) which has a Schunk PG70 parallel jaw gripper. This gripper was simulated using the *Gripper Sensor* and *Gripper Actuator* plugins (see Fig. 3.8). Each finger of the gripper has a Weiss Robotics sensor (DSA 9205) attached, which were modeled with the *Tactile Sensor* plugin explain in the following section.

Tombatossals Robot

The most completed platform that has been modelled is the humanoid torso *Tombatossals* (see Fig. 3.9) at the Robotics Intelligence Lab at Universitat Jaume I. It is the only platform that has been modelled completely using dynamics (see Sect. 4.6).

(a) (b) (c) (d) (e)

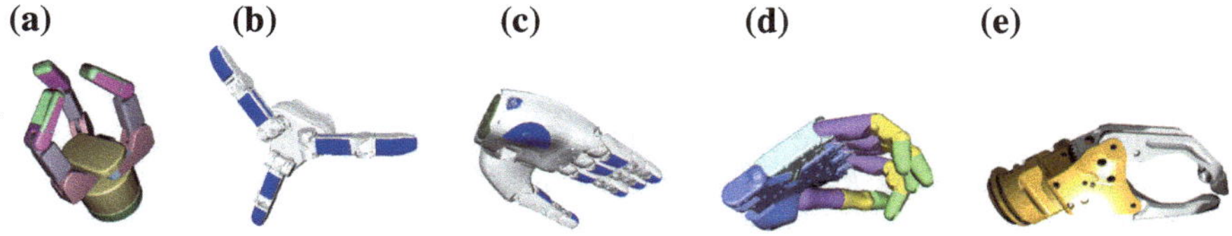

Fig. 3.7 Different robot models generated with the Robot Editor (ray-traced images). **a** the Barrett hand **b** the Schunk hand SDH **c** the Schunk hand SAH **d** the Shadow hand, **e** A myoelectric upper extremity prostheses of Otto Bock

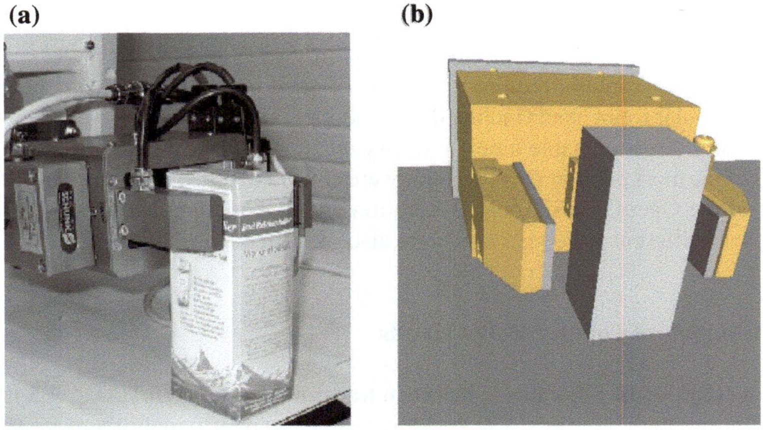

Fig. 3.8 Schunk PG70 gripper at LUT. **a** Real **b** Simulated

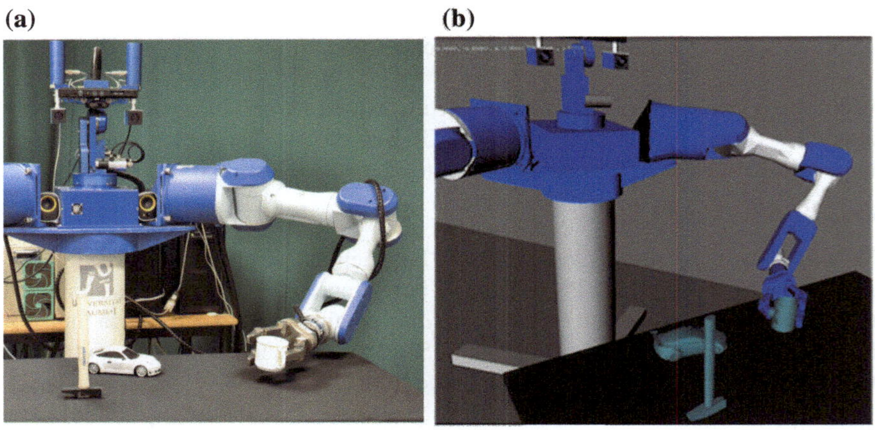

Fig. 3.9 The humanoid robot Tombatossals. **a** Real **b** Simulated

It has 25 DOF and is composed of two 7 DOF Mitsubishi PA10 arms. The right arm has a 4 DOF Barrett Hand and the left arm has a 7 DOF Schunk Dexterous Hand 2.0 (SDH).[8] Both hands are endowed with a Weiss Tactile Sensor system on the fingertips. Each arm has a JR3 Force-Torque sensor attached on the wrist between the arm and the hand. The visual system is composed of a TO40 4 DOF pan-tilt-verge head with two Imaging Source DFK 31BF03-Z2 cameras. Attached to the centre of the pan-tilt unit there is a $Kinect^{TM}$ sensor from $MicrosoftCorp.$

The torso and the head CAD models were created according to the measurements of the real robot. The models of the Mitsubishi PA10 arms were taken from the

[8] http://www.schunk.com

OpenRAVE robot model database and modified to fit the exact manipulator model (PA10 7CE). The model for the Barrett hand was also obtained from OpenRAVE. The Schunk SDH hand CAD model was provided by the manufacturer but the masses of each part were not established, so they were estimated for each piece from the geometry and the weight of the whole hand. This hand has three fingers each with two joints. An additional degree of freedom executes a coupled rotation of two fingers around their roll axis. Each finger is padded with two tactile sensor arrays: one on the distal phalanges with $6 \times 13 - 10$ cells and one on the proximal phalanges with 6×14 cells. The tactile sensors were simulated using a tactile sensor plugin developed by the authors described in Sect. 3.4. The force sensor was also simulated as the plugin for OpenRAVE. The cameras were modelled using the camera sensors provided in OpenRAVE with the same parameters and position as the real cameras.

The angular motors, available in the robot arm and hand joints, have been simulated with the ODE controller provided by OpenRAVE. Each simulated servo-motor was parametrized by the maximum speed, the maximum acceleration and the maximum torque that the motor can apply.

The humanoid robot ARMAR

Another platform that has been modeled is the humanoid robot ARMAR-IIIa[9] [16] consisting of seven subsystems: head, left arm, right arm, left hand, right hand, torso, and a mobile platform (see Fig. 3.10). The head has seven DOF and is equipped with two eyes, which have a common tilt and an independent pan. For the visual perception of the environment, the humanoid's active head features two stereo camera systems, one with a wide-angle lenses for peripheral vision and one with a narrow-angle lenses for foveal vision. For grasping and manipulation, the robot provides a 3 DOF torso and two arms with 7 DOF each. The arms follow an anthropomorphic design: 3 DOF for each shoulder, 2 DOF in each elbow and 2 DOF in each wrist. Each arm

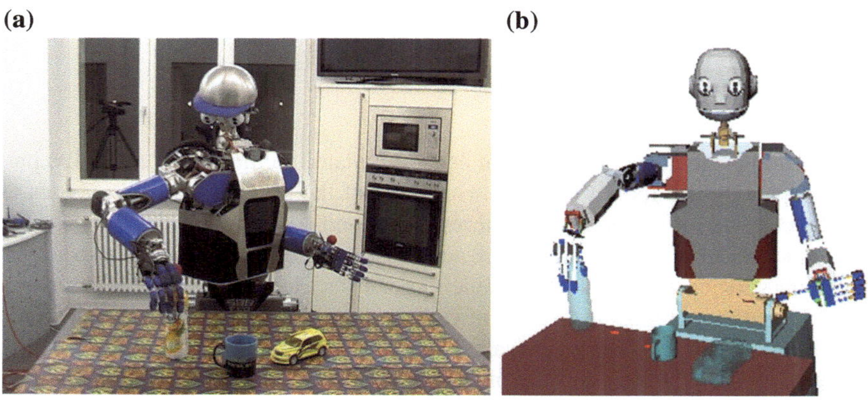

Fig. 3.10 The humanoid robot ARMAR-IIIa. **a** Real **b** Simulated

[9] http://his.anthropomatik.kit.edu/english/241.php

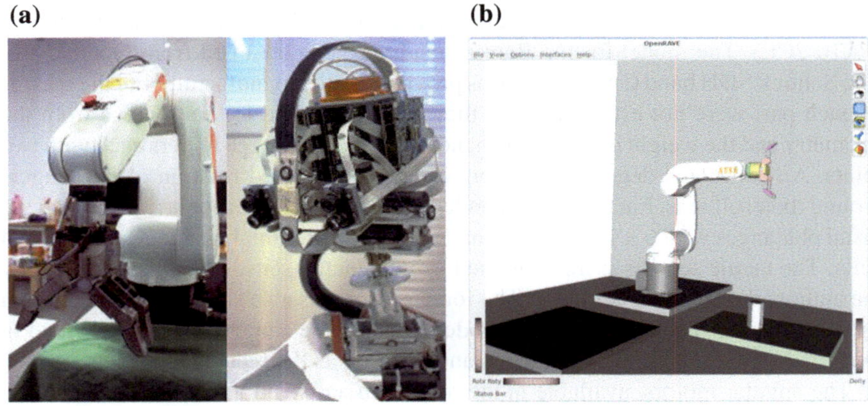

Fig. 3.11 Schunk hand SDH and the Kuka KR5 sixx R850 arm. **a** Real **b** Simulated

is equipped with a pneumatic-actuated five-fingered hand. The robot moves on a wheeled holonomic platform. The geometry and kinematics of this robot platform were modelled using the Robot Editor.

KTH platform

The platform used at the Royal Institute of Technology (KTH) (see Fig. 3.11) consists of the Karlsruhe active head [17] equipped with two stereo cameras, a peripheral (wide-field) and a foveal (narrow-field) one. The robotic head has 7 DoF. Five of these are used for controlling the viewing direction while the remaining two mainly vary the vergence angle between the left and right camera systems, thereby enabling fixation on objects. As a manipulator, we use a 6 DoF Kuka KR5 sixx R850 arm[10] that is equipped with a three-fingered Schunk SDH.

3.4 Simulated Tactile Sensor

This section presents the development of the simulated tactile sensor created as part as the OpenGRASP toolkit [18, 19] with some of the same physical properties as a real tactile sensor has, such as compressibility and friction. In order to create the sensor model three different areas were addressed: tactile sensor model construction, modelling soft contacts and friction modelling. The tactile sensor is created based on a geometry patch enabling the creation of various shape tactile sensors. A contact force model was created that enables the calculation of surface forces as well as the holding torque around the contact surface and the stick-slip phenomenon. The model does not include the load spreading to adjacent texels generated by the material thickness. Also manufacturing and mechanical imperfections are ignored at this point.

10 http://www.kuka-robotics.com

In order to test the proposed model, different experiments were conducted. First, the physical properties of the simulated tactile sensor were validated using static and dynamic tests. Also an experiment on simulated robot grasping vs. real robot grasping was carried out by a robot hand grasping an object and by the corresponding model on the simulator performing the same actions. The use of the sensor to create a complete dynamic simulation of a humanoid robot is presented in the next chapter. Finally, the sensor model potential was explored by applying it to modelling human touch.

3.4.1 Related Work

Touch, combined with vision, are the main senses that allow humans to perform dexterous manipulation. For this reason, sensors that can retrieve tactile information have been developed in order to equip robot hands with such a sense.

Tactile sensors are defined as devices that can measure different properties of an object through physical contact between the sensor and the object [20]. They can measure mechanical properties including pressure, normal and shear forces, torques, slip and vibrations, or other properties like temperature or moisture. In robotic manipulation, only the mechanical properties of the contact are studied, typically sensing normal forces and contact positions. Different methods exist for constructing tactile sensors (for a review see [21, 22]). There are tactile sensors based on various principles such as resistive, capacitive, optical, ultrasonic, magnetic or piezo-eletric sensors.

The performance of the real tactile sensors developed until now is far from human sensing capabilities. Nevertheless, they have been used in robot manipulation in the last few years for different purposes including reactive robot control [23–25], collision detection [26, 27] and object recognition [28]. These different applications of tactile sensors show the importance of their use in robot manipulation. Having a tactile sensor model that enables tactile sensing simulation and, more generally, complete simulation of robot grasping, will be of great benefit to the robotics community.

In general, robot grasping simulations have traditionally been using kinematics instead of dynamics. This is due to the fact that robot dynamic simulation is a very challenging problem. The most common simulation method for robot grasping has been the impulse method (GraspIt!, ODE, Bullet, etc.). The impulse method is a very effective method in terms of calculation speed for simulating structures that form open kinematic chains which robotic manipulators usually are. The drawback of using impulse methods for solving the constraints between the bodies is that the accuracy of the constraints is dependent on the mass ratio of the two objects. This means that if the robotic manipulator has very light grippers attached to a heavy wrist the joints connecting the bodies can suffer from instability. The same applies to impulse based collisions. Using a penalty method this mass dependency can be avoided in the collisions but the solution becomes sensitive to variables, such as the time-step, due to the stiffness of the system.

Simulation of tactile sensors for robot grasping is a fairly new field of research. Using the available simulation environments and physics engines, there are some existing models of tactile sensors. [29] presented a model using GraspIt! which provide contact and force feedback. However the problems mentioned by using the impulse methods to model dynamics limited the simulation. Also the use of rigid contact cannot model the deformation of a compliant fingertip or holding torque around the contact. There has been some attempts to use penalty methods for rigid body simulations as [30–32], but they used only Coulomb friction which leads to errors at near zero sliding velocities. [33] used a point spread function to model the response of the tactile sensors and characterize them via robotic experiments. However the effect of friction is ignored as well as the actual sensor values.

In the following section, a tactile sensor model is presented which overcome most of the weakness of the previous tactile sensor simulations.

3.4.2 Tactile Sensor Model

The purpose of this work was to make a simulation model of a tactile sensor, not just by emulating the function but by modelling the actual physical properties starting from the formation of an actual contact patch to including a full friction description.

Different tactile sensors are available with a variety of shapes. Rigid sensors range from the simple planar sensors to ones shaped to curve around a robot fingertip. Also some flexible sensor types are available which can be for example wrapped around a humanoid robot arm. The idea was to create a tactile sensor that could be adapted to model any shape. For this reason, the tactile sensor is created based on a geometry defined by a triangularized mesh, which can be obtained from a CAD model.

The tactile sensor model is designed to measure the object mechanical properties, such as contact forces and contact positions. Acquiring the force enacting on the tactile sensor from an existing simulation environment can be troublesome. For example, if the simulation environment uses a non-penetrating collision method the conversion from a collision force to tactile data leads to extensive assumptions. A non-penetrating collision method will not give reliable results on force build up as the objects are not allowed to interpenetrate. Even if the objects are allowed to interpenetrate, the collision will change the object velocity in a single time-step making the tactile data unreliable.

The solution proposed is to calculate the contact forces within the tactile sensor by using a soft contact method and then apply them to the bodies. This means disabling all other collisions between the two bodies making the tactile force the only force acting between the two bodies. That way the pressure detected by the sensor and the pressure applied to the body are identical. Also this will ensure that the tactile sensor can acquire all the related data for producing tactile information as it is all calculated within it.

The chosen method consists of using a contact patch with several texels in order to form the tactile sensor. Single texels are used in order to determine collisions against other objects and to calculate the resulting collision forces. This provides an accurate and fast solution for solving the collision equations. Currently the number of contact points is equal to the number of texels in the tactile sensor but in the future a single texel can be modified to include several contact points in order to further increase the resolution of the sensor.

The contact forces are calculated on each contact point and are used by the simulator to grasp the object and for the tactile sensor feedback.

3.4.2.1 Geometry-Based Tactile Sensor

The simulated tactile sensor can be formed based on a triangularized geometry. This was done so that differently shaped sensors could be easily defined. In Fig. 3.12 different variations of a tactile sensor geometries are presented: a simple grid, a spherical surface and a geometry modelling a human fingertip. The arrows represent the normal directions of the different triangles.

The texels of the simulated tactile sensor are constructed using the vertices from the sensor geometry. For example, in the case of a planar tactile sensor consisting of an array with 8 rows and 6 columns, one would draw a 7×5 grid having 8×6 vertices to represent the centers of the tactile cells (see Fig. 3.13).

For each vertex, the normalized sum of all normals of the triangles connected to it is calculated and used as a normal direction to the sensor element. The sensor element's maximum penetration needs to be defined in the sensor parameters. It is used to place the beginning of a vector pointing in the normal direction to the vertex (see Fig. 3.14). This vector in turn is used to calculate the intersection against all possible triangularized target geometries. The forces calculated at this point are explained in the following section.

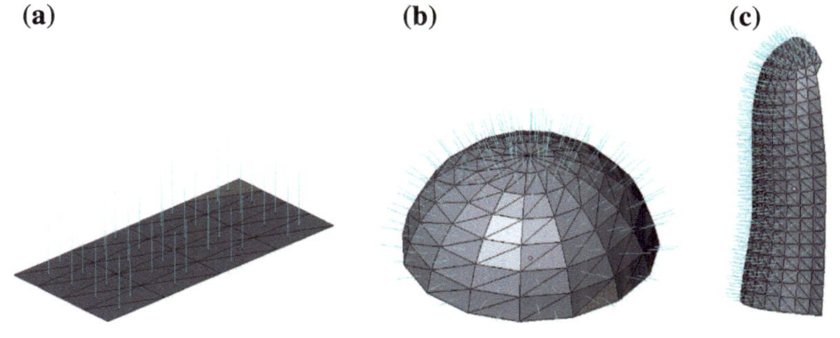

Fig. 3.12 Example of tactile sensor geometries: **a** a simple grid **b** a spherical surface and **c** a human fingertip

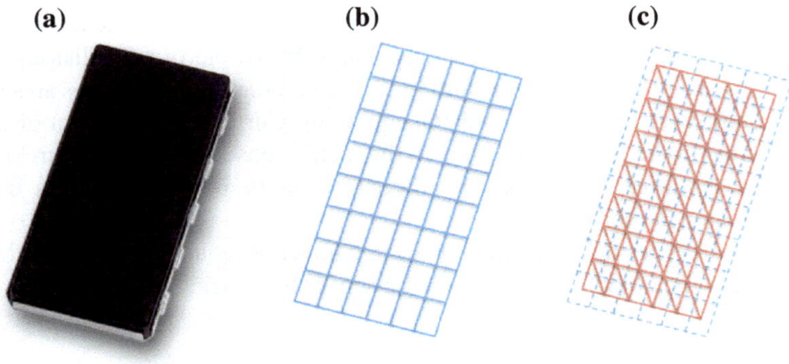

Fig. 3.13 Example of a simulated tactile sensor construction: **a** real tactile sensor, **b** geometry of the sensor and **c** simulated tactile sensor elements

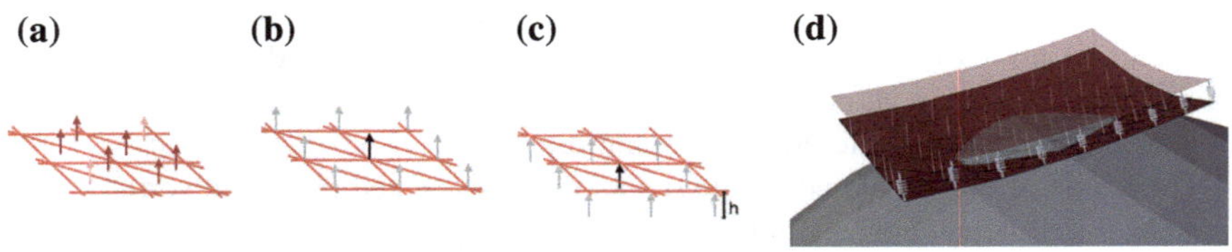

Fig. 3.14 Determining collision vector for each vertex: **a** Triangle normals, **b** Normalized sum of triangles connected for each vertex **c** vertex vectors placed at the maximum penetration (h) and **d** overview of the sensor in touching an object and the contact area

3.4.3 Contact Force Model

When a collision between the sensor and an object occurs, the contact information (position, relative velocity, penetration, etc.) is used to calculate the force in a single texel. Contact forces are described using the soft contact approach which allows small penetration between contacting bodies taking into account local deformations. The amount of this penetration is calculated accordingly with the maximum penetration defined for each sensor. A brief description is presented here but the complete set of equations used can be found in [18].

On each contact point, the contact force (F_C) can be written as:

$$F_C = F_n + F_t \tag{3.1}$$

where F_n is the normal force produced by the soft contact and F_t is the tangential force represented by friction. The contact force F_n in the normal direction of the plane can be calculated as a linear spring-damper element. The tangential friction force F_t can be evaluated using the LuGre friction model [34] which accounts for both static and sliding phenomena based on a bristle deflection interpretation. It captures the dynamic behavior of the contact surface using the first order differential equation for bristle deflections.

The forces and torques can be calculated on each contact point. These forces are applied to the body where the tactile sensor is attached as well as to the body that the tactile sensor is colliding with. They are also used to retrieve sensor feedback information.

A summary of the steps needed to construct the simulated tactile sensor and to calculate the contact forces are outlined in Algorithm 1.

Algorithm 1: How to construct a simulated tactile sensor and obtain the sensor data

---

Initialization:
Using a mesh representing the real sensor geometry, create the simulated sensor texels;
Parametrize the sensor elements with the maximum penetration and the values needed to calculate the normal force (k, c_n) and friction (μ_s, μ_d, σ_0, σ_1, c_t);
Result: Tactile Sensor Readings

```
begin
    for each vertex of the simulated sensor element do
        Place a vector pointing to it, with a magnitude equal to the maximum penetration and
        with a direction equal to the sum of normals of the triangles connected to it;
    for each time-step do
        for each vertex do
            Calculate the intersection of the vector with the target objects;
            if they are in collision then
                Create a contact point on the intersection;
        for each contact point do
            Calculate the contact force in the normal direction; Calculate the tangential
            friction; Add these two components to get the total contact force;
        for the sensor's body and each target body do
            Calculate and apply the forces and torques;
        Convert the forces to tactile values;
```

3.4.4 Tactile Sensor Plugin

The simulated tactile sensor plugin has been developed and is available in OpenGRASP. The development of the tactile sensor plugin included the definition of the tactile sensor geometry, the tactile sensor data and the implementation of the sensor interface specified by OpenRAVE.

The tactile sensor requires a set of parameters to be specified for each sensor. The model presented on this study allows the creation of sensors based on any geometry which is specified by the mesh of the body to which the sensor is attached. For each sensor, the parameters necessary to calculate the contact forces and its thickness need to be defined.

The tactile sensor data is the structure returned by the tactile sensor when it is requested. It contains the size of the tactile array, a vector with the tactile values calculated in each cell and the sum of all the tactile values. This structure is the same as the one used by the real tactile sensor, which makes the real and simulated sensors feedback appear identical to the controllers.

The plugin is an implementation of the sensor interface. OpenRAVE creates a new tactile sensor when specified on a robot definition. Each time step, when the sensor is updated, it gets the positions and velocities of the sensor and objects and checks if they are colliding. On each contact point, it calculates the contact forces and torques are calculated and they are applied to the sensor and objects. Finally, given that a real tactile sensor produce tactile intensities instead of forces, these intensities can be calculated using a linear conversion and the tactile data structure is updated. Controllers can query and use this tactile sensor feedback as required.

3.4.5 Experiments on Robot Grasping

In order to determine the simulated tactile sensor performance, different experiments were carried out. They can be divided into two groups: tests for validating the physical properties and features of the tactile sensor and tests for applying the model to situations where the real sensor is commonly used. In the following sections these experiments are described in detail.

All the experiments were carried out with a simulated model of the Schunk PG70 parallel jaw gripper or by using a single finger model of the Schunk gripper. Each finger of the gripper had a tactile sensor attached to it. This is a resistive tactile sensor, with an array of cells of 14 rows by 6 columns.

3.4.5.1 Validation of the Physical Properties of the Simulated Tactile Sensor

The basic physical properties of the tactile sensor element were tested by two different static methods. First, the capability of the sensor to return correct force feedback is validated. Second, the capability of the sensor to bear loads in static situations was tested. This is important as the sensor needs to be able to hold the objects for an extended period of time. This shows that the proposed method does not suffer from the same problem as some other models with near zero velocity.

Force sensor feedback

The load sensing capability of the sensor was tested by loading the sensor with a varying weight object and the total force from the tactile elements was then compared to the total force required to keep the object static (see Fig. 3.15a). The weight test range was from 0.1 to 10 kg. The parameters used in this simulation were a spring constant (k) of 1,000 N/m and a damping coefficient (c_n) of 10 Ns/m, with a time-step size of 0.0001 s. Similar results can be acquired with smaller and greater masses by

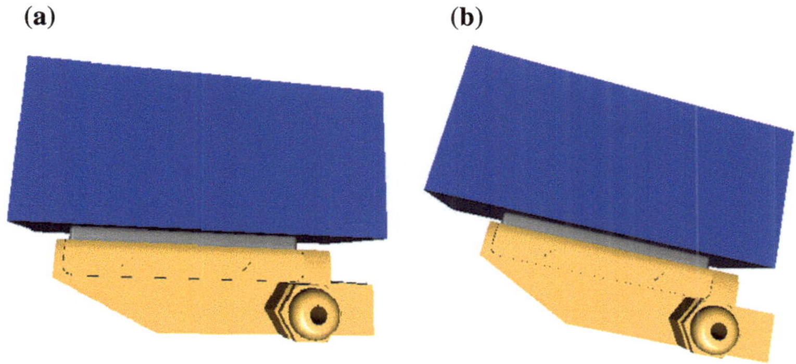

Fig. 3.15 Simulated Robot finger conducting static tests

Table 3.2 Tactile weight sensing capability

Object weight (kg)	Required force [N]	Tactile force feedback [N]
0.1	0.9800	0.9800
1	9.8000	9.7999
10	98.0000	97.9899

changing the k and c_n values of the sensor in order to keep the cube horizontal and stable.

A summary of the tests for 3 different weights is presented in Table 3.2. The table shows the object weight, the theoretically required force to hold the object static and the sum of the forces obtained by each texel from the tactile sensor. Given that in this case all the forces are pointing to the same direction, we can calculate the total force as the algebraic sum of the force values. The sum of the forces does not exactly match the required force due to the slight tilting of the object. Therefore a fraction of the load is carried by the friction force which is not included in the normal force returned. Based on the results, the tactile sensor give out correct force readings.

Static friction bearing capability

One of the main problems for collision models is the static holding friction. Most of the collision response methods will gradually let the object slip away from the grasp as near zero sliding velocity is handled poorly by the friction model [35]. The static friction bearing capability was tested by tilting the finger by 10 degrees and setting the 1 kg cube on top of it as shown in Fig. 3.15a and seeing if the friction will hold the object static.

In Fig. 3.16 the results of the test are shown. The creep value is the position difference to the initial position. The cube moves slightly from the starting position and then settles in position due to friction. The cube settles to the same position within 10 decimal accuracy within 10 s. The simulation was continued for 20 min confirming that the cube does not move from that position.

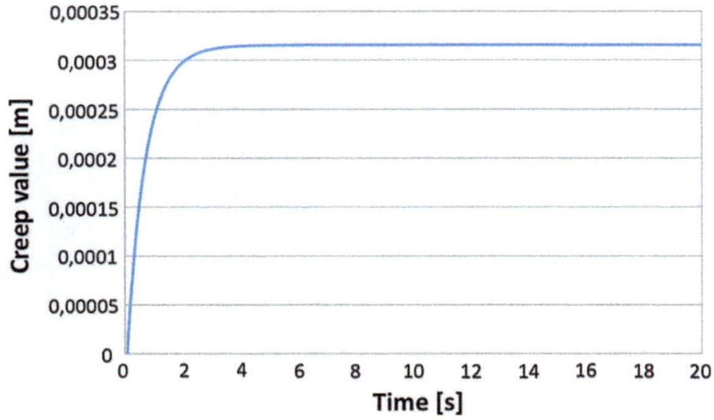

Fig. 3.16 Creep value as position difference to initial position [m] of the cube when placed on a slanted tactile

This proves that the tactile sensor is capable of holding loads at a static position and the importance to include a proper friction model to a grasping simulation.

3.4.5.2 Testing the Model on Robot Grasping Applications

Grasping various shape objects

Tactile sensors are normally used to grasp various shape objects. Therefore it is important to verify that the tactile sensor model can also handle grasping variously shaped objects. Four different objects were chosen representing typical graspable objects such as: a sphere, a box, a cylinder and a more complex form like a spray bottle. The objects were placed between two robot fingers and the tactile image and contact forces were observed during a simulation run. In Fig. 3.17 the object, the corresponding mesh, a grasping position the corresponding tactile images and the normal contact forces are shown. It can be seen that the tactile sensor is able to detect as well as hold on to all of the shapes. The tactile resolution can be changed in the simulation model by increasing the density of the mesh used to create the sensor.

It can also be seen in Fig. 3.17 that the simulated tactile sensor gives out highly accurate impression imprints of the grasped objects. This tactile impressions can be used in a variety of application such as object recognition using tactile exploration and grasp stability evaluation [27].

Comparing tactile readings with the real sensors

A task of grasping and picking up a cube was selected as the test scenario. The idea was to perform the same task using this robot and compare the results with the ones obtained by executing the same actions in the simulator. In order to accomplish this, a high level controller was implemented using an abstraction architecture presented in [36], which allows to switch between real and simulated hardware transparent from the controller point of view.

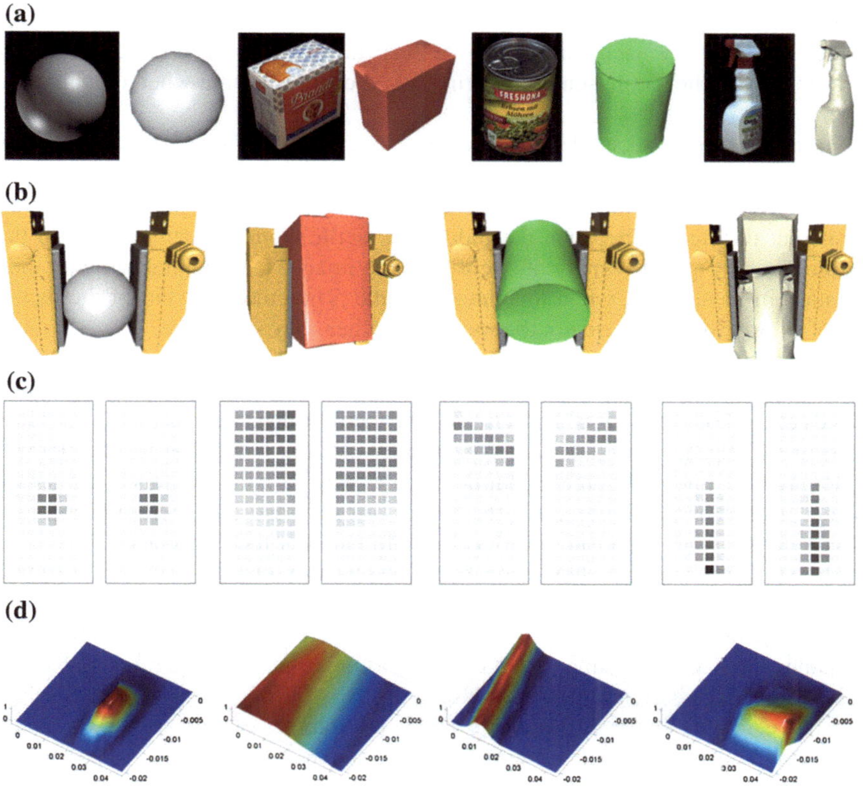

Fig. 3.17 Testing results of grasping four different objects. **a** Real object picture and simulated object **b** Snapshots of robot fingers and objects at the grasping position **c** Tactile images taken at the grasping position (*white* = 0, *black* = max) for the left and right tactile sensor **d** Normal force readings in Newtons at each texel position measured in the grasp position for the right tactile sensor

The tactile sensor feedback was used to control the grasping force and to determine the stability of the grasp. The experiment was defined as an abstract action consisting of five primitive actions: approach, grasp, lift, move down and release. The controller turns this abstract information into the gripper specific primitives and transitions. It then drives the Schunk actuator using velocity control until the first contact with the tactile sensors is detected. After the initial touch the controller switches to force control by setting the maximum current of the Schunk gripper based on the tactile sensor feedback. The tactile value sums are used as the reference for the force control. This particular case shows the function of the tactile sensor in combination with the robot controller. Given the abstraction architecture's ability to be embodiment independent, the same controller was used to control the real robot as well as the simulated case.

An image sequence of the real robot performing the work cycle with the tactile images produced by each sensor, can be seen in Fig. 3.18. The experiment was performed with the tactile sensor covering 30 percent of its area when touching the cube.

The actions were executed as expected but closer investigation of the tactile values revealed some problems. When grasping the cube, even pressure was applied to each tactile sensor that should return, as a result, very similar tactile images. However, significant differences between the individual tactile elements can be seen in the bottom row of Fig. 3.18 where one sensor tactile image is substantially lighter than the other in most cases, when touching the cube. Therefore the sum of the tactile values on opposite sides are not the same even when they should be.

The simulated robot performed the same work cycle as the real one. Images of the robot on each work cycle phase are shown in Fig. 3.19 with the corresponding tactile images at the bottom. Closer inspection of the tactile values reveals that the sensors perform exactly as expected. Under even pressure, the individual tactile element values are the same in the area touching the cube. The individual texels in the sensor are not visible in the figure since the values are identical and therefore no color difference is visible. Opposite sensors also return similar tactile values.

The simulation model shows consistent results whereas the real tactile sensor results vary on each work cycle. This is due to the fact that in simulation there are no manufacturing flaws or problems from wearing. The system consistently performs the same way under the same conditions. In addition, the detection tolerances from individual elements do not pose a problem.

When using the tactile imprints to evaluate the stability of a grasp, the simulated tactile could be used to teach a controller what a stable grasp looks like. The sensors currently available, will fail to detect a stable grasp based on the simulated images due to difference between the simulated and the real sensor feedback given the imperfections between the real individual sensors. The simulated tactile could be used to teach the controller what an ideal stable imprint would look like, but the real

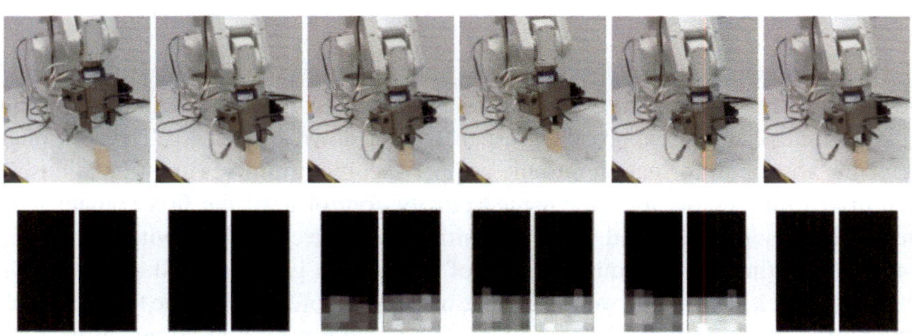

Fig. 3.18 Real robot performing the chosen work cycle. On the *top row*, pictures of the robot performing the task on each stage. On the *bottom*, the tactile images generated by left and right sensors (*black* = 0, *white* = max)

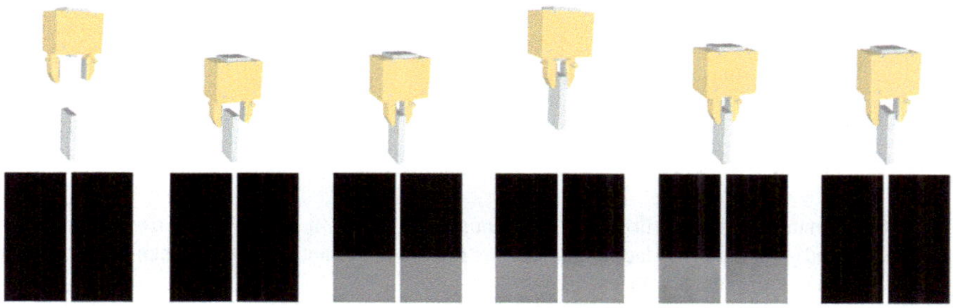

Fig. 3.19 Simulated robot performing the chosen work cycle. On the *top row*, pictures of the simulated robot performing the task on each stage. On the *bottom*, the tactile images generated by left and right sensors. (individual texels not visible due to close to identical values)

sensor would not display an identical image. An algorithm for allowing deviations from the ideal imprint would have to be developed.

There are some features such as resolution, noise, hysteresis, creep and aging that characterize a tactile sensor [37] and that should be considered in simulation in order to get a more similar response to the real sensors readings. The resolution in simulation can be changed freely, so it can match the spatial and temporal resolution of the real tactile sensor which are hardware dependent. From the real sensor's results it can also be seen that the tactile values suffer from noise in the results whereas, at the moment, the simulated sensor reports the changes in the force directly without any interferences. The hysteresis in the real tactile can also be significant due to the material covering the tactile elements. Finally, the real sensor material also causes some creep in the results as the foam cover resistance changes over time even under constant pressure. This change settles after some time but there is always some creep even after an extended period. The cover material also causes aging to be a problem when using the real tactile.

All these features of the real tactile sensor can be added to the tactile sensor model. The difficulty is that the variance from sensor to sensor can be quite considerable as shown in Fig. 3.18 as well as in [33].

3.4.5.3 Mimicking the Human Finger with a Tactile Sensor

An experiment with the human hand was conducted to test the versatility of the tactile model as well as to show the direction of the future work. The tactile sensor were created using geometries of human fingers (human hand model courtesy of the Biomechanics and Ergonomics Group, Universitat Jaume I) and then a cylinder shaped object was placed in between the index finger and the thumb (see Fig. 3.20). In the image it can be seen that the finger senses the object in a correct manner. The purpose of the experiment is to show that any shape can be used to form the tactile sensor. This is not an extensive test as such, but gives ideas for various uses of the

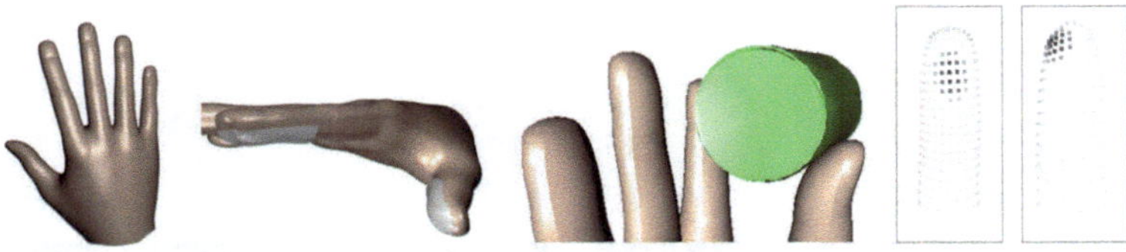

Fig. 3.20 Human finger tactile holding a cylinder shaped object. It can be seeing from *left to right*: the human hand geometry, the tactile geometries for the index and thumb, the hand grasping the cylinder and the tactile images for each finger

simulated tactile sensor for the future. The freedom of form of the tactile sensor element can be extremely useful in modelling different tactile applications.

Real tactile sensors are not accurate enough to model the human touch just yet. This sensor model can give researchers a head start on researching human touch given that the simulated sensors do not suffer from the same restrictions as the actual real tactile sensors do.

3.4.6 Discussion

The experimental results of the tactile sensor model showed good performance in being able to produce tactile feedback. Problems arised when trying to calibrate the tactile model to correspond exactly to a certain real tactile sensor. This is due to the variations in the real tactile values which makes the process extremely difficult.

However, the use of this tactile sensor model may enable researchers to do experiments that should be theoretically possible but, due to the current limitations in the existing hardware, are still difficult.

The collision detection method for solving the contact points is currently a brute force method. Improvements on this area can greatly improve the overall computation times of the tactile sensor model. The collision response method is computationally effective. The calculations are straight forward and the dimensions of the equations of motion matrices will not be increased given that the penalty forces are applied to the dynamic model as external forces. Future work includes improving the collision detection times and improving computational efficiency.

3.5 Conclusion

In this chapter we have presented a fully operational simulation toolkit for robot grasping and manipulation. Its main design principles are extensibility, interoperability and public availability. In its development we have used existing and widely-available components to ensure its standardization and easy adoption. We have also emphasised providing additional tools and features that provide users with a quick start to enhance utility through features such as the robot editor based on Blender, COLLADA file format, two Physics Abstraction Layers (PAL and FISICAS), and models of existing robot hands.

Additionally, a simulation model of a tactile sensor was presented. The simulation model is based on soft contact modelling with a full friction description. The sensor was tested using the most commonly use cases of tactile sensors in robotic grasping. The simulated tactile sensor performed all the tests including stable grasping without any errors and can be used as a tactile sensor model. The model can be updated to behave in the same manner as a specified type of tactile sensor such as one from Weiss Robotics. This would entail modifying the stiffness as well as adding delay and load spreading that are due to the covering material and other electrical properties such as noise.

References

1. Barsalou, L.W.: Grounded cognition. Annu. Rev. Psychol. **59**(1), 617–645 (2008), http://www.annualreviews.org/doi/abs/10.1146/annurev.psych.59.103006.093639, pMID: 17705682
2. Gerkey, B., Vaughan, R.T., Howard, A.: The player/stage project: tools for multi-robot and distributed sensor systems. In: 11th International Conference on Advanced Robotics (ICAR 2003). pp. 317–323. Coimbra, Portugal, (June 2003)
3. Koenig, N., Howard, A.: Design and use paradigms for gazebo, an open-source multi-robot simulator. In: IEEE International Conference on Intelligent Robots and Systems. vol. 3, pp. 2149–2154 (2004)
4. Carpin, S., Lewis, M., Wang, J., Balakirsky, S., Scrapper, C.: Usarsim: a robot simulator for research and education. In: Robotics and Automation, IEEE International Conference on. pp. 1400–1405 (2007)
5. Jackson, J.: Microsoft robotics studio: a technical introduction. Rob. Autom. Mag. IEEE **14**(4), 82–87 (2007)
6. Kanehiro, F., Fujiwara, K., Kajita, S., Yokoi, K., Kaneko, K., Hirukawa, H.: Open architecture humanoid robotics platform. In: Proceedings of the IEEE International Conference on Robotics and Automation. pp. 24–30 (2002)
7. Michel, O.: Cyberbotics ltd. webots tm : professional mobile robot simulation. Int. J. Adv. Rob. Syst. **1**, 39–42 (2004)
8. Miller, A., Allen, P.: Graspit! a versatile simulator for robotic grasping. Rob. Autom. Mag. IEEE **11**(4), 110–122 (2004)
9. Malvezzi, M., Gioioso, G., Salvietti, G., Prattichizzo, D., Bicchi, A.: Syngrasp: a matlab toolbox for grasp analysis of human and robotic hands. In: IEEE International Conference on Robotics and Automation (ICRA2013) (2013)
10. Diankov, R.: Automated Construction of Robotic Manipulation Programs. Ph.D. thesis, Carnegie Mellon University, Robotics Institute, (Aug 2010)
11. Nguyen, B., Trinkle, J.: dvc3d: a three dimensional physical simulation tool for rigid bodies with intermittent contact and coulomb friction. In: First Joint International Conference on Multibody System Dynamics, (May 2010)
12. Drumwright, E., Hsu, J., Koenig, N., Shell, D.: Extending open dynamics engine for robotics simulation. In: Ando, N., Balakirsky, S., Hemker, T., Reggiani, M., von Stryk, O. (eds.) Simulation, Modeling, and Programming for Autonomous Robots, Lecture Notes in Computer Science, vol. 6472, pp. 38–50. Springer, Berlin (2010)
13. Boeing, A., Bräunl, T.: Evaluation of real-time physics simulation systems. In: Proceedings of the 5th international conference on Computer graphics and interactive techniques in Australia

and Southeast Asia. pp. 281–288. GRAPHITE '07, ACM, New York, USA (2007) http://doi.acm.org/10.1145/1321261.1321312
14. Burns, D., Osfield, R.: Open scene graph a: Introduction, b: Examples and applications. Virtual Reality Conference, IEEE 0, 265 (2004), http://www.computer.org/portal/web/csdl/doi/10.1109/VR.2004.10021
15. Gaiser, I., Schulz, S., Kargov, A., Klosek, H., Bierbaum, A., Pylatiuk, C., Werner, R.O.T., Asfour, T., Bretthauer, G., Dillmann, R.: A new anthropomorphic robotic hand. In: Proceedings IEEE/RAS International Conference on Humanoid Robots (HUMANOIDS). pp. 418–422 (2008)
16. Asfour, T., Regenstein, K., Azad, P., Schröder, J., Vahrenkamp, N., Dillmann, R.: Armar-iii: An integrated humanoid platform for sensory-motor control. In: IEEE/RAS International Conference on Humanoid Robots. pp. 169–175 (2006)
17. Asfour, T., Welke, K., Azad, P., Ude, A., Dillmann, R.: The Karlsruhe Humanoid Head. In: IEEE/RAS International Conference on Humanoid Robots. Daejeon, Korea, (Dec 2008)
18. Moisio, S., Leon, B., Korkealaakso, P., Morales, A.: Simulation of tactile sensors using soft contacts for robot grasping applications. In: 2012 IEEE International Conference on Robotics and Automation (ICRA). pp. 5037–5043. Saint Paul, Minnesota, (May 2012)
19. Moisio, S., León, B., Korkealaakso, P., Morales, A.: Model of tactile sensors using soft contacts and its application in robot grasping simulation. Rob. Auton. Syst. 61(1), 1–12 (2013), http://www.sciencedirect.com/science/article/pii/S0921889012001911
20. Lee, M.H., Nicholls, H.R.: Tactile sensing for mechatronics : a state of the art survey. Mechatronics **9**(1), 1–31 (1999)
21. Dahiya, R., Metta, G., Valle, M., Sandini, G.: Tactile sensing - from humans to humanoids. IEEE Trans. Rob. **26**(1), 1–20 (2010)
22. Tegin, J., Wikander, J.: Tactile sensing in intelligent robotic manipulation–a review. Int. J. Ind. Rob. **32**(1), 64–70 (2005)
23. Felip, J., Morales, A.: Robust sensor-based grasp primitive for a three-finger robot hand. In: IEEE/RSJ International Conference on Intelligent Robots and Systems, 2009. IROS 2009. pp. 1811–1816, 10–15 (2009)
24. Kragic, D., Crinier, S., Brunn, D., Christensen, H.: Vision and tactile sensing for real world tasks. In: IEEE International Conference on Robotics and Automation. vol. 1, pp. 1545–1550, 14–19 2003
25. Prats, M., Sanz, P.J., Pobil, A.P.: Reliable non-prehensile door opening through the combination of vision, tactile and force feedback. Auton. Rob. **29**(2), 201–218 (2010)
26. Hsiao, K., Chitta, S., Ciocarlie, M., Jones, E.G.: Contact-reactive grasping of objects with partial shape information. In: IEEE/RSJ International Conference on Intelligent Robots and Systems, (Oct 2010)
27. Bekiroglu, Y., Laaksonen, J., Jorgensen, J., Kyrki, V., Kragic, D.: Learning grasp stability based on haptic data. In: In Proceedings of the RSS 2010 Workshop: Representations for object grasping and manipulation in single and dual arm tasks, (June 2010)
28. Okamura, A.M., Cutkosky, M.R.: Feature Detection for Haptic Exploration with Robotic Fingers. Int. J. Rob. Res. **20**(12), 925–938 (2001)
29. Tegin, J., Wikander, J.: Simulating tactile sensors in robotic grasping. In: Proceedings of The Third Swedish Workshop on Autonomous Robotics. vol. 1. Stockholm, Sweden, Sept 2005
30. Hasegawa, Y., Shikida, M., Shimizu, T., Miyaji, T., Sasaki, H., Sato, K., Itoigawa, K.: Micromachined active tactile sensor for hardness detection. Sens. Actuators A: Phys. **114**(2–3), 141–146 (2004), selected papers from Transducers 03
31. Ghafoor, A., Dai, J.S., Duffy, J.: Stiffness modeling of the soft-finger contact in robotic grasping. J. Mech. Des. **126**(4), 464–656 (2004)
32. Ho, V.A., Hirai, S.: Three-dimensional modeling and simulation of the sliding motion of a soft fingertip with friction, focusing on stick-slip transition. In: IEEE International Conference on Robotics and Automation. pp. 5233–5239 (2011)
33. Pezzementi, Z., Jantho, E., Estrade, L., Hager, G.: Characterization and simulation of tactile sensors. In: Haptics Symposium, 2010 IEEE. pp. 199–205 (2010)

34. de Wit, Canudas: C., Olsson, H., Astrom, K., Lischinsky, P.: A new model for control of systems with friction. IEEE Trans. Autom. Control **40**(3), 419–425 (1995)
35. Drumwright, W., Shell, D.: An evaluation of methods for modeling contact in multibody simulation. In: IEEE International Conference on Robotics and Automation, (May 2011)
36. Laaksonen, J., Felip, J., Morales, A., Kyrki, V.: Embodiment independent manipulation through action abstraction. In: IEEE International Conference on Robotics and Automation. pp. 2113–2118 (3–7 2010)
37. Schneider, A., Sturm, J., Stachniss, C., Reisert, M., Burkhardt, H., Burgard, W.: Object identification with tactile sensors using bag-of-features. In: IEEE/RSJ International Conference on Intelligent Robots and Systems, 2009. IROS 2009. pp. 243–248, 10–15 (2009)

Chapter 4
Applications of Robot Grasping Simulation

4.1 Introduction

In robotics research, simulators have an important role in the development and demonstration of algorithms and techniques in areas such as path planning, grasp planning, mobile robot navigation, and others. There are several reasons for using robot simulations. First, they allow exhaustive testing and tuning of mechanisms and algorithms under varying environmental conditions. Second, they avoid the use, or misuse, and wearing of complex and expensive robot systems. Also, the environment and its attributes can be completely controlled. Additionally, simulation software is cheaper than real hardware so that a large number of experiments can be efficiently performed.

Specifically in robot grasping, the simulator is used to perform different grasps before the real robot makes an attempt. In theory, there are an infinite number of candidates grasps applicable to an object. Grasp synthesis algorithms select a set of relevant grasp hypothesis among all the candidate grasps. These algorithms can be generally divided into analytical and empirical approaches [1]. Analytical approaches determine the contact locations on the object and the hand configuration that satisfy task requirements through kinematic and dynamic formulations. They are usually formulated as a constrained optimization problem over different criteria that measure dexterity, equilibrium, stability or dynamic behaviour. Empirical approaches, on the other hand, rely on sampling many grasp candidates for an object, generate these in simulation, an then evaluate their quality according to a specific metric.

In this chapter, the use of the simulation for selecting relevant grasp hypothesis using empirical grasp synthesis algorithms is demonstrated. These approaches differ in the amount of prior information the robot is assumed to have about the object and in how the inferred grasps are executed. The approaches can be divided into grasping known, unknown or familiar objects according with the taxonomy proposed by [2]:

B. León et al., *From Robot to Human Grasping Simulation*,
Cognitive Systems Monographs 19, DOI: 10.1007/978-3-319-01833-1_4,

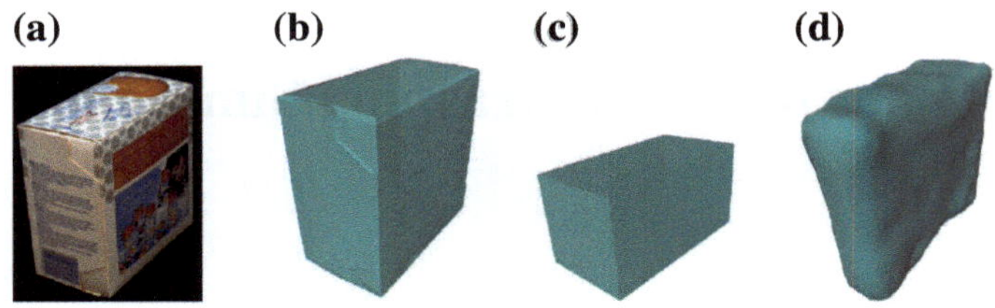

Fig. 4.1 Grasping approaches in simulation. **a** Real object. **b** Simulated known object. **c** Simulated familiar object. **d** Simulated unknown object

- *Known objects*: When known objects are used, the robot has an accurate geometrical model of the object. They can be created off-line using different technologies, like laser scans. Then, the grasping problem consists on object recognition and pose estimation, and object-specific grasps can be retrieved from a database.
- Unknown objects In the case of unknown objects, the object is new and previously unseen by the robot, therefore an approximate model has to be created on-line usually using heuristics (commonly object shape features in relation to the robot hand) to generate and rank grasp.
- Familiar objects Grasping familiar objects deal with the problem of transferring grasp experience from previously encountered objects to a similar new object. The underlying assumption is that new objects, similar to the old ones, can be grasped in a similar way. Objects can be similar in terms of shape, color or texture. Also, the similarity can be based in terms of more high-level features, such as the functionality that they afford or the possession of similar parts or categories.

All these approaches are developed and evaluated in simulation. An example of the object model for each approach is shown in Fig. 4.1. Using the simulator to replicate the kinematics of the real platform, we show how it can be used to select an appropriate grasp and plan its trajectory in order to successfully grasp objects that are known, unknown or familiar to the robot.

Grasping known objects using a simulator is a well established technique. Therefore, in the first section, we present the existing approach used by OpenRAVE—the chosen simulation framework—when the knowledge of the grasping objects is assumed to be known.

However, this is an assumption that is not realistic if the goal is to achieve autonomous robot grasping. First, the robot should be able to grasp any object present in the environment, therefore it will be impossible to construct an object database with the model of all possible objects that it may encounter. Additionally, the models of the objects are normally represented as triangle meshes obtained using high-precision scanners techniques which are normally not available on mobile platforms. As the goal is to robustly identify the models and their configurations (position, orientation) in the environment of a robot in order to apply the grasp knowledge from the models on the real objects, there are several factors such as noise, uncertainty, occlusions,

Table 4.1 Prior information the robot is assumed to have for each application

Objects	Approach	Prior information			
		Exact object model	Object database	Grasp database	Task
Known	Using uncertainty metric MOOM	Yes	Yes	Yes/No	No
Unknown	Using symmetry assumptions	No	No	No	No
Familiar	Using task constraints	No	Yes	Yes	Yes

exact model of the given grasping object, an existing object database, an existing precomputed database of stable grasps and the manipulation task to be performed after grasping

missing parts due to lack of texture and restricted viewpoint that make this a very difficult task.

Therefore, we have proposed three different independent approaches varying the amount of prior information the robot is assumed to have, in order to infer grasps under more complex environments. The prior information for each application is shown in Table 4.1.

In the fist application, we present an approach to grasp known objects improving the existing ones by incorporating the depth-sensor information into the grasp planing, which enables to select better grasp hypotheses and also to constraint approach directions and the path planning space, thus reducing the amount of time needed to find a successful grasp hypotheses. In order to do that we propose MOOM: an overlap metric between the recognized model and the actual part of the object that is seen from the camera.

The second application shows a technique proposed to grasp unknown objects predicting their full shape by making use of the observation that many, especially man-made objects, possess one or more symmetries. The aim is to show that even if the object model is not an exact representation of reality, it is close enough to enable the simulator to try different grasp alternatives and select an appropriate one.

In the third application, we start with the assumption that the object is a familiar object, meaning that it is geometrically similar to a known model. The objective of this approach is to achieve autonomous grasping of objects according to their category and a given task. We show how to incorporate the object category and the task to transfer task-specific grasp experience between objects of the same category. The effectiveness of the approach is demonstrated on two humanoid robot platforms.

Finally, we present the last application with a different focus. In the previous ones, we have used kinematics instead of dynamics given that taking into account masses, forces, inertias, static and dynamic frictions, and even elasticities and deformations, is a very challenging problem, specially when it comes to considering the interactions between contacting bodies. A large number of parameters, which are difficult to determine, affect the dynamics behaviour of the involved parts. In this last application, we want to demonstrate how we can achieve a complete dynamic simulation of a humanoid robot using the developed framework.

The applications shown in this chapter are the result of my collaboration with several researches and resulted in the following publications: [3–5]. In the following sections the complete approaches are presented but my role in them and focus was the development of the prediction block using OpenGRASP and its interfaces to integrate it with the rest of the grasping pipeline modules.

4.2 Grasping Known Objects: Existing Approaches

There are many challenging problems addressed by the robotics community currently studied in simulation in which the knowledge of the complete world model or of specific objects is assumed to be known. Simulators focusing in solving manipulation problems such as GraspIt! or OpenRAVE can be used to solve these problems. Given that OpenRAVE was chosen as the simulation framework for this work (see Sect. 3.3), its approach to grasping known objects [6] is presented in this section and later compared with the new approaches developed in this book in each application.

The simulator requires the following parameters to describe a manipulation task: an object to be grasped considered as the *target* object, the goal positions to move the target/robot to, any target or robot motion constraints and regions that should be avoided. Once a robot has found the target object, using knowledge of physics and the object geometry, the simulator can automatically build models of how to stably grasp the object and manipulate it. Combining the grasping models and robot kinematics with the task constraints defined by the problem specification, the space of all possible robot goals and movements that will satisfy the task can be computed.

A *grasp* is parametrized by (Fig. 4.2):

- *6D end-effector pose*: transformation of the robot hand in the target's coordinate system
- *Grasp preshape*: starting finger configuration (joint angles)
- *Approach direction*: direction vector describing the direction in which the robot hand approaches the target, in the target's coordinate system
- *Stand-off*: distance along negative direction of the approach direction to get out of collision
- *Roll*: wrist rotation angle around the approach direction axis.

The information required to perform a manipulation task can be divided into information independent of the current state of the environment and information obtained at run-time like obstacles and target positions that the robot has no way of pre-computing beforehand. Therefore, OpenRAVE offers several off-line algorithms that can pre-compute the first type of information into a form that makes on-line retrieval as quick as possible. This information is stored into several databases:

- Inverse Kinematic database: Analytic inverse kinematics equations for each arm required for exact computation of the arm reachability space that is used for many

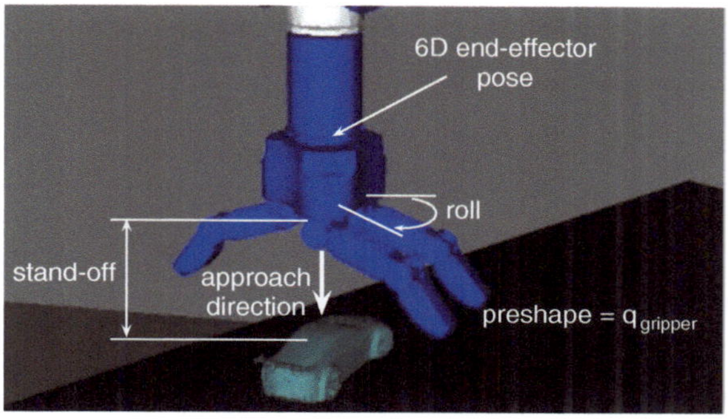

Fig. 4.2 Parameters defining a grasp

planning heuristics. It uses the *ikfast* algorithm for finding the most robust closed-form solutions.

- *Grasp Hypothesis database*: From the target object perspective, OpenRAVE generates all the grasps that may handle the object according to the task specification.
- *Convex decomposition database*: provides much simpler geometries to work with.

These databases are optimized for fast usage times during on-line planning. However, computational time when generating them is not a key factor given that these databases can be computed off-line and they can be generated faster with a cluster of computers. They are stored using unique hashes for the robot and target bodies so that the data is indexed more consistently.

4.2.1 Grasp Hypothesis Database

The Grasp hypothesis database is presented here in more detail, given that it is the most involved in grasp simulation and has been used for different applications described later in this chapter.

A representation is needed in order to save the space of valid grasps. There have been different grasping models proposed to provide such a representation, including grasp sets or parametrized models like Support Vector Machines or explicit goal regions. Diankov [6] claims that valid grasps spaces should be expressed as sets, given that although it might be more compact to explicitly parametrize the space of good grasp, such methods can be susceptible to modeling errors.

In order to create the grasp database, the grasp set is generated with the combination of the different parameters specified that define a grasp:

- A set of gripper pre-shapes, stand-offs and rolls that want to be used to grasp the object can be specified.
- Approach directions can also be individually specified or generated automatically using the following procedure: the surface of the object is sampled by casting

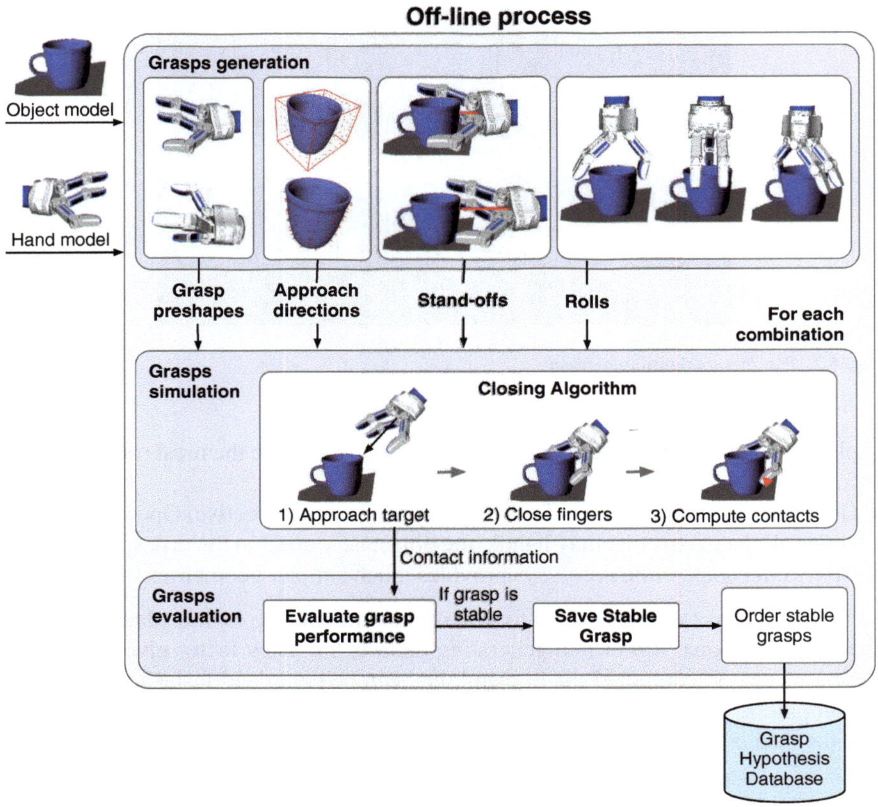

Fig. 4.3 Generation of grasp hypothesis database

rays from a bounding box to the target object. Then for each hit point, the surface normal of the object is computed. Using this normal, it is possible to sample all directions that are within a fixed angle between each normal.

The process of generating the grasp hypothesis database is depicted on Fig. 4.3. It starts parameterizing each grasp with a combination of the input parameters.

One ray is defined on the hand's coordinate system along the direction of the palm and another ray is defined on the target object towards the approach direction specified. The robot hand pose is computed by aligning these two rays together and make the directions point towards each other. Then, a common grasping strategy is used approaching the gripper to the object using this direction and close the fingers until they cannot move anymore [7]. Tactile or force sensors are not considered given that if the fingers can constantly provide torque, the point contact analysis can be enough for simulating the grasping of rigid objects with a rigid hand. The contact points with the object are extracted and the performance of that contact needs to

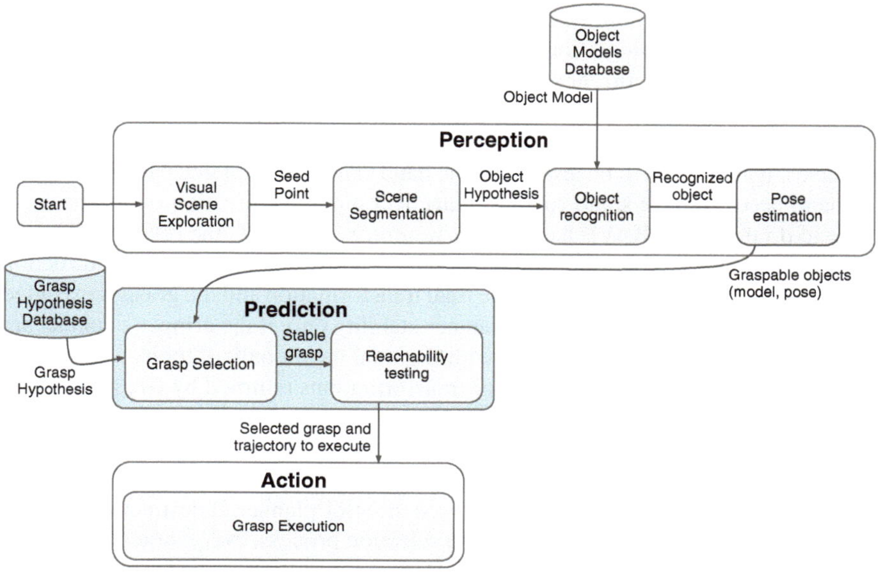

Fig. 4.4 Grasp pipeline for grasping known objects

be evaluated to determine if the grasp is stable. This performance can be measured using different strategies. The one most used is to measure grasp stability using force closure (see Definition 2.8) to check if any force acting on the object can be compensated by the forces generated on the contact points. However other measures can be used, which have been reviewed in Chap. 6. The grasps that are stable are saved on the database. When all the combinations are tested, the set of stable grasps are ordered by priority.

4.2.2 OpenRAVE Grasping Pipeline

An overview of the grasp cycle for grasping known objects is shown in Fig. 4.4. It is subdivided into three major building blocks. Given visual input from cameras, the perception block is responsible for detecting an object, recognizing it and estimating its pose. The prediction block takes in the object pose and model and selects an appropriate grasp from the Grasp Hypothesis Database that has to be reachable in the current robot configuration. The selected grasp is then executed on the robotic platform.

This pipeline is taken as the base grasp cycle and, in each of the applications shown later in this chapter, we present the alterations made. We will focus on the prediction block—highlighted in blue—as it represents the main role of the simulator, and we will also describe its integration with other modules.

After receiving the current state of the environment with the target object model and its estimated pose, the simulator needs to select an stable grasp and validate it in the current environment. The grasp first needs to be collision-free from other obstacles, it also needs to be tested that the hand can successful grasp the object, and the hand needs to have small room to maneuver along its direction of approach. In order to do this, OpenRAVE has a *GraspValidator* which sets the robot hand to the preshape and checks environment collisions as it moves along its negative direction of approach. Then the hand is set to the final transformation and the grasp is executed in simulation. Because the grasp guarantees stability only if the gripper's contact are from the target object, all collisions with the final hand configurations are checked that are only with the target. The set of transformations returned by *GraspValidator* are all checked for inverse kinematics solutions and that the entire robot is in the free space. If there is a grasp transformation that does not have a solution, then the robot will not be able to complete the task and the grasp is skipped. Otherwise, the transformation is taken as the goal to feed the RRT planner. If destination goals of the target are considered during the grasp selection process, every target destination is checked for possible collisions with the target object, gripper and arm.

This grasp cycle is used each time that a robot needs to select a valid grasp to be successfully executed by the robot platform. This approach using OpenRAVE has been extensively used and validated [8–15]. In the following sections, we present how we take this approach and apply it to more complex grasping tasks in order to grasp unknown and familiar objects.

4.3 Grasping Known Objects: Using Uncertainty Metric MOOM

This application is based on the known object approach, also known as *Model-driven grasping approach*, which assume that a model of the objects exist. The models of the objects are normally represented as triangle meshes obtained using high-precision scanner techniques which are normally not available on mobile platforms. The goal is to robustly identify the models and their configurations (position, orientation) in the environment of a robot in order to apply the grasp knowledge from the models on the real objects. Noise, uncertainty, occlusions, missing parts due to lack of texture and restricted viewpoint make this a very difficult task.

We propose the Model-Object Overlap Metric (MOOM) between the recognized model and the actual part of the object that is seen from the camera. The MOOM constrains the grasp hypotheses computed on the full model by giving a lower weight to those hypotheses that use unseen parts of the object. We show as well how the MOOM can be used not only to select better grasp hypotheses but also to constrain approach directions and the path planning space, thus reducing the amount of time needed to find a successful grasp hypotheses without time consuming involving reachability checks and collision free paths.

With the advent of the Microsoft Kinect providing dense 3D information, some promising methods have been lately presented dealing with the problem of recognizing objects in the world using 3D synthetic meshes as input for a training stage [16, 17]. The recognition results can be used to guide a grasp planner to perform operations on real objects using the precomputed grasp knowledge.

Still, the grasp planner assumes that the recognized model matches the reality although we have no real evidence of that due to occlusions (hidden parts by the object itself or by other elements in the environment) and moreover, the recognition results could be wrong or undetermined. In order to better deal with these uncertainties, we propose to effectively use the evidence of data to constrain the grasp hypotheses and avoid those with less probability of being correct.

The work presented in this section is the result of a collaboration with Aitor Aldoma,[1] Walter Wohlkinger and Javier Felip.[2] The advantages of the presented method are three-fold:

1. It allows to deal with uncertainties of non-observed data but still makes use of the assumptions taken by object recognition, which is less restrictive than symmetry [3] or similar assumptions.
2. For on-line grasp planners, the MOOM can be used to speed up the computation of successful hypotheses by pruning approaching directions to the object that intersect parts of the model with lower MOOM.
3. Finally, it provides a metric to rank those grasp hypotheses with force-closure according to online observations. Because grasp hypotheses are labeled as successful or unsuccessful depending on force-closure, there is no real criteria to decide which successful hypotheses should be actually executed, apart from reachability and collision checks.

In the remaining of this section, we review similar model-driven grasping methods and an object recognition technique to provide the pose of the 3D models in the actual scene. We present how the MOOM is computed and how it is integrated in the grasping pipeline. Finally, we present an experimental evaluation to demonstrate how the metric can be used to improve speed and grasp accuracy on an online grasp planner provided by OpenRAVE.

4.3.1 Related Work

Model-driven grasping is a well studied problem in the literature. There have been several approaches relying on object recognition or similarity between objects to

[1] Vision4Robotics Group, Automation and Control Institute, Vienna University of Technology, Vienna, Austria.

[2] Robotic Intelligence Lab, Universitat Jaume I, Castellon, Spain.

transfer grasps between objects represented as 3D meshes [18] or from 3D mesh models to real objects [19, 20].

More recently, there have been some interest in combining different strategies for grasping. In [13], the authors use a probabilistic model to decide which is the best object representation that can be used to grasp a real object: fitting primitives to the partial point cloud obtained with the depth sensor, recognizing the object against a database of objects and by estimating their pose applying known grasps or directly computing grasp hypotheses in the mesh reconstructed from the partial view.

Brook et al. [21] present a similar approach to [13]. Instead of deciding which is the best model representation to grasp the actual object, the method tries to reach a consensus between all different model representations and the best grasp is selected. The first object representation is a recognition-based representation obtained using a matching algorithm based on a 2-dimensional matcher that iteratively aligns each model in the database to the segmented cluster of the depth sensor. Because of the simple recognition technique used, only objects rotationally symmetrical or standing in a known orientation can be recognized. The second object representation is based solely on the segmented clusters obtained from the depth sensor and the grasp hypotheses are computed using a set of heuristics.

There are several limitations in both [13] and [21]: (1) the set of heuristics used by the grasp planner are not easy to port to more dexterous hands than the PR2 gripper and (2) the recognition method is only able to recognize objects with very specific constraints and does not scale with the number of objects. It is as well not trivial to decide which representation should be ultimately used when no consensus is reached between the different representations.

Therefore, we propose to strongly rely on the model representation obtained using more advanced recognition methods but transfer the absence or presence of real observations to the model representation so that the grasp planner can implicitly use it.

4.3.2 MOOM: Model-Object Overlap Metric

In previous approaches, after recognition and pose estimation, the grasping pipeline assumes that the recognized model matches the reality although we have no real evidence of that, apart from the metrics obtained from the recognition module.

Why should the grasping pipeline throw away the valuable information from the depth sensor after recognition? The main idea behind the computation is to encode the real evidence of data in the mesh representation together with the recognized object and pose so that all available information is provided to the grasping pipeline to use it wisely.

The computation of the overlap metric is straightforward. Given a mesh $\mathcal{M}$ aligned to a point cloud $\mathcal{P}$ representing the object—both in camera coordinates—the overlap metric is computed by building an octree from $\mathcal{P}$ and finding the closest neighbour for each vertex v_i of $\mathcal{M}$ in the octree representation. Given $p_i \in \mathcal{P}$ is the closest

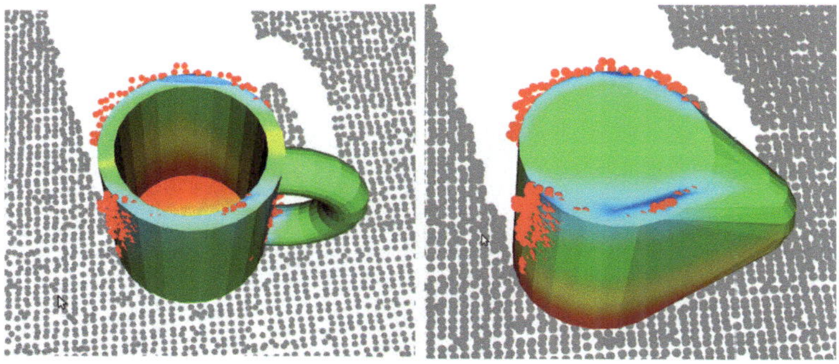

Fig. 4.5 A closer look to a recognized mug. The colors of the mesh represent the overlap between the segmented cluster (*red dots*) and the mesh. The color representation of the overlap increasing from *red* (*worse overlap*) to *blue* (*best overlap*). **a** Colored object mesh ($\mathcal{O}_{mesh}$). **b** Colored convex hull ($\mathcal{O}_{hull}$)

neighbour of $v_i \in \mathcal{M}$, the overlap metric at v_i is computed as follows:

$$O(v_i, p_i) = \frac{1}{1 + \frac{||v_i - p_i||}{M_d}} \qquad (4.1)$$

where M_d represents the maximum distance considered for a point v_i to overlap. We use a value of 2 mm for M_d in all experiments. The overlap metric inside the mesh triangles is obtained by linear interpolation between the 3 vertices of the triangle although it is not stored and always computed on the fly when needed.

We have used this overlap information at two different levels: (1) to sort the approach directions so that those intersecting at points of the model with high MOOM are preferred and (2) to check the quality of the force-closure grasps by computing the overall MOOM of the contact points between the manipulator and the object.

Two meshes are used for this purpose: the complete object mesh and its convex hull. Let $\mathcal{M}$ be the mesh aligned with the point cloud $\mathcal{P}$ and $\mathcal{C}_{\mathcal{M}}$ the convex hull of $\mathcal{M}$. For both, $\mathcal{M}$ and $\mathcal{C}_{\mathcal{M}}$ the overlap metric with $\mathcal{P}$ is computed obtaining two colored meshes, $\mathcal{O}_{mesh}$ and $\mathcal{O}_{hull}$ respectively. $\mathcal{O}_{hull}$ is used to filter and sort the approach vectors and $\mathcal{O}_{mesh}$ to check the quality of the grasp. An example of the overlap metric for a recognized mug and its convex hull are shown in Fig. 4.5.

4.3.2.1 Overlap Metric for Approach Vectors MOOM($\mathcal{O}_{hull}$, p)

The overlap metric MOOM is used to find directions to approach the manipulator to the object in areas with less uncertainty. Let p represent the contact point between the object and the approach vector. The MOOM($\mathcal{O}_{hull}$, p) is computed to sort the

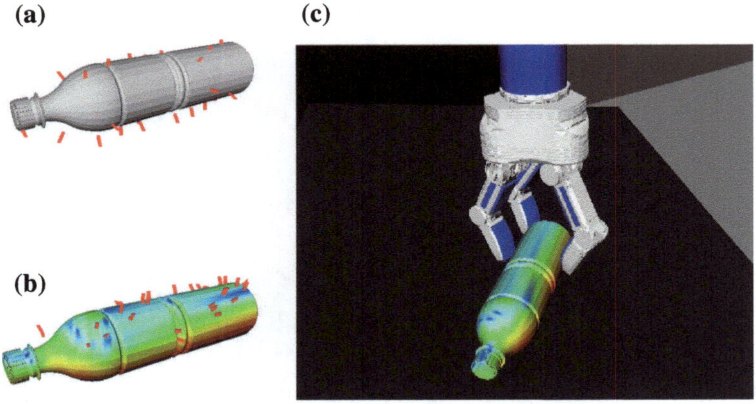

Fig. 4.6 White bottle lying on the table. Without MOOM there is no implicit criteria to decide which approach vectors should be tried first. **a** First 30 approach vectors retrieved by OpneRAVE without MOOM. **b** First 30 approach vectors sorted using MOOM. **c** First force-closure grasp hypothesis found with a MOOM($\mathcal{O}_{mesh}$, P) higher than 0.5

approach directions so that those intersecting at points of the model with high MOOM are preferred.

To sort the approach directions, $\mathcal{O}_{hull}$ is used to ensure that the algorithm will not discard those approach vectors coming from the top in non-convex objects like mugs or bowls. Additionally, to account for the table plane obstacle, each vertex $v_i \in \mathcal{O}_{hull}$ is reweighted according to its distance to the table plane by a factor f_i computed as follows:

$$f_i = \begin{cases} 1 & d(v_i, t_p) \geq min_d \\ \frac{d(v_i, t_p)}{min_d} & \text{otherwise} \end{cases} \tag{4.2}$$

where t_p represents the table plane, $d(v_i, t_p)$ the point-to-plane distance from point v_i to the plane t_p and min_d the minimal distance to the table needed by the embodiment to approach an object parallel to the table. In our case, we use a value of 10 cm for objects higher than 10 cm, otherwise the maximum height of the object. Modifying the overlap metric in this way, we enforce the planner algorithm to first test those approach vectors that are more likely to provide a collision free path when trying to reach the object (see Fig. 4.6).

4.3.2.2 Overlap Metric for Grasp Hypotheses MOOM($\mathcal{O}_{mesh}$, P)

Once the grasp is simulated, the algorithm computes the overall overlap metric MOOM of the contact points, which assign better rates to grasps whose contact points are located on object areas with high overlap. Let P represent all contact points between the object and the manipulator. The MOOM($\mathcal{O}_{mesh}$, P) is computed

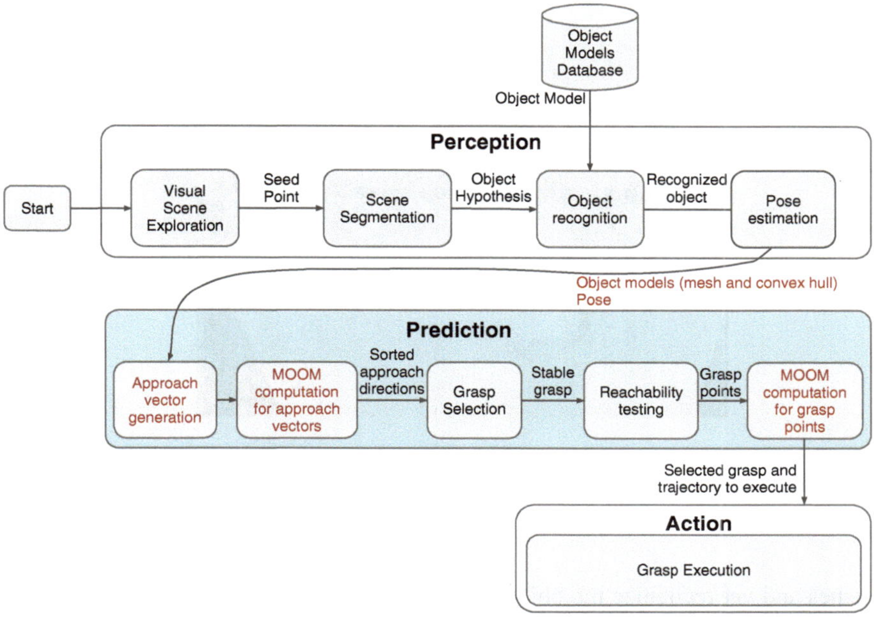

Fig. 4.7 Grasp pipeline for grasping known objects using MOOM. Alterations of the referenced pipeline (Fig. 4.4) are shown in *red*

by averaging the $\mathcal{O}_{mesh}(p_i)$ where $p_i \in P$ resulting in a metric to rank the overall quality of the grasp in relation to the observed data (see bottom of Fig. 4.6).

4.3.3 Grasping Pipeline

An overview of the grasp cycle is shown in Fig. 4.7. Given visual input from cameras, the perception block is responsible for detecting an object, matching it to a known model and estimating its pose. The prediction block takes in the object pose and model and selects an appropriate grasp with high MOOM, reachable in the current robot configuration. The selected grasp is then executed on the robotic platform.

4.3.3.1 Perception

Model-driven grasp approaches rely on object recognition and pose estimation to apply grasps learned on a mesh to a real object. Grasp planners usually work with triangle mesh representations of the objects to compute force closure grasp hypotheses and therefore, we seek an object recognition method that can be trained on triangle

Fig. 4.8 RGB-D image from the Kinect and recognition results overlayed. The mesh is overlayed on the point cloud and colored according to the overlap between the segmented cluster and the mesh

meshes and yet recognize the objects obtained with a depth sensor like the Kinect. This eases training as there is no need for several representations of the objects: both perception and prediction pipelines can use the same models. Figure 4.8 show an image of a example scene together with the reconstructed point cloud and the recognition results overlayed.

We decide to use the Clustered Viewpoint Feature Histogram (CVFH) descriptor and the recognition pipeline presented in [16] which has been shown to perform well in similar scenarios. CVFH is a semi-global, view based descriptor composed by several histograms based on angular normal distributions of the object surface. Even though, we decided for this specific recognition pipeline, the rest of the approach applies to any recognition pipeline able to match mesh representations to the real world.

The mesh representation of the objects used for the experiments have been obtained from different sources: simple objects like boxes or cylinders have been manually modeled, other objects were obtained from the KIT Object Database [22] and others were automatically classified using the Shape Distribution on Voxel Surfaces descriptor (SDVS [17]) and its scale was obtained using the pose alignment method presented in [23]. Models obtained using the automatic method do not match the real object completely, making the recognition and grasping task even more challenging. The decision of using models that do not match real objects completely is motivated by: (1) it eases training as there is no need to the real objects to build a mesh representation of it that can be used by the grasp planner and (2) it allows to show how the overlap metric can be used to enhance grasping even in these challenging scenarios where the 6DOF pose estimation will never be accurate due to differences between model and real object.

4.3.3.2 Prediction

We present two alternatives of how the overlap metric can be used depending if a grasp hypothesis database is available or not.

Overlap metric without a grasp database. We use the grasping pipeline from OpenRAVE. It includes a planner algorithm called fastGrasping which purpose is to find the first feasible grasp for an object as fast as possible without generating a grasp database. The approach vectors are generated taking the bounding box of the object and sample its surface uniformly. The first intersection of the object and a ray originating from each point going inwards is taken and the normal of the object's surface at the intersection points is taken as the approaching direction of the manipulator. These approach vectors are used to generate a set of grasps, using a combination of predefined preshapes, stand-offs and wrist rolls. Each one of these grasps is tested first for force-closure and second to be reachable by the robot manipulator in a trajectory free of collisions with the environment. The first grasp to fulfill these conditions is sent for execution to the robot.

We have modified this algorithm to include the overlap information to sort the approach directions so that those intersecting at points of the model with high MOOM are preferred and to check the quality of the force-closure grasps by computing the overall MOOM of the contact points between the manipulator and the object (see Algorithm 1).

Algorithm 1: OpenRAVE's fast grasping with MOOM

```
input  : O_hull, O_mesh, robot
output: An executable grasp
begin
    AV: = get approach vectors for O_hull;
    for each av in AV do
        compute MOOM(O_hull, p) (see Sect. 4.3.2);
    sort (AV); //in decreasing order;
    for each approach vector do
        test grasping object in the approach vector direction;
        if grasp is "force closure" then
            if grasp is reachable and collision free then
                compute MOOM(O_mesh, P) for grasp contact points;
                if MOOM(O_mesh, P) ≥ threshold then
                    send grasp for execution;
                    break;
```

Overlap metric with a grasp database. Algorithm 2 presents a variation of the first algorithm when an off-line generated grasp database is available. It can be seen that this kind of model-driven approaches can also take advantage of MOOM to increase efficiency and robustness.

Algorithm 2: Grasping with MOOM using a grasp database

input : $\mathcal{O}_{hull}$, $\mathcal{O}_{mesh}$, robot, DB:= Database of force closure grasp for the given object and robot
output: An executable grasp
begin
  for *each grasp in* DB **do**
    compute MOOM($\mathcal{O}_{hull}$, p) for approach vectors (see Sect.4.3.2);
    compute MOOM($\mathcal{O}_{mesh}$, P) for grasp contact points;
  $DB_{filtered}$:= filter(DB) for which MOOM($\mathcal{O}_{hull}$, p) $\geq$ threshold;
  sort ($DB_{filtered}$) by MOOM($\mathcal{O}_{mesh}$, P); //in decreasing order;
  for *each grasp in* $DB_{filtered}$ **do**
    if *grasp is reachable and collision free* **then**
      send grasp for execution;
      break;

4.3.4 Experimental Setup and Evaluation

In order to test and validate the method presented, we have compared the *fast grasping* grasp planner from OpenRAVE with Algorithm 1. The plans have been executed on a real robotic platform to validate whether the planned grasps were successful. The selected platform in this case is the anthropomorphic torso *Tombatossals* (see Sect. 3.3.5).

In the environment where the experiments are performed there is a table in front of the robot that acts as a surface where the objects stand. The object set is composed of several household objects of different shapes: boxes, cups, mugs, bowls, bottles, duct tapes and fruits (Fig. 4.9). Although the grasp pipeline allows more than one object at a time, to perform the grasping experiments, one single object is put on the table in front of the robot in any position and orientation. The object position has to be close enough to allow the robotic arm to reach and grasp the object. The presence of a dominant plane in the scene is assumed to segment the object of interest.

The following metrics are captured for the different grasp planners for the evaluation in order to measure improvement in both accuracy and computational efficiency:

- *Grasp hypotheses tested*: Percentage of grasp hypotheses checked on the simulator until the first valid grasp is sent for execution to the real platform from all possible grasp hypothesis generated by the planner. The total number of grasp hypotheses is a combination of approach vectors, stand-offs, wrist rolls and preshapes.
- *Time*: Elapsed time by the grasp planner until a grasp hypothesis is sent to the robot.
- *Grasp success*: Whether the final grasp execution was successful or not on the real robot. This input is given by the operator: a grasp is considered successful if the robot is able to lift the object, failure otherwise.

Fig. 4.9 Object set used for the experiments. The objects are identified with their position in the grid from *top* to *bottom* and *left* to *right*. This identifier is used on the experiment graphs (starting at 0)

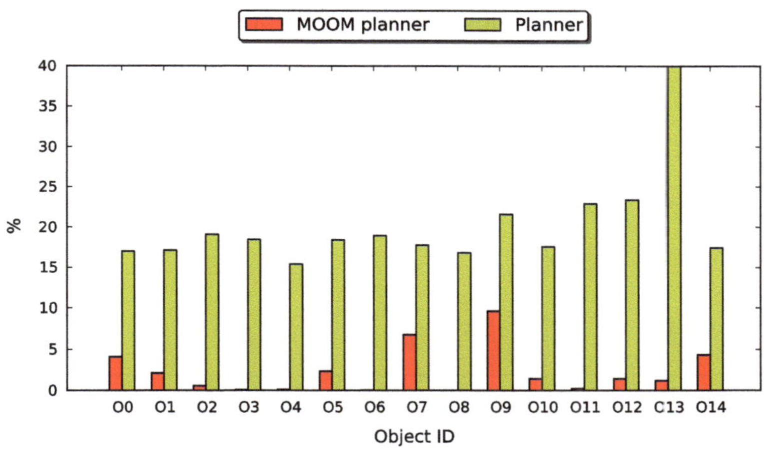

Fig. 4.10 Percentages of grasp hypotheses tested until the first feasible grasp can be sent for execution

Figure 4.10 presents the percentage of grasp hypotheses tested from all generated grasp hypotheses until a feasible one is found that can be sent for execution. It is clear from the figure that the goodness of the grasp hypotheses using MOOM is much higher than those from the standard planner which are randomly selected. Because grasp hypotheses are generated from the set of approach vectors, the approach vectors sorted using MOOM are found on collision free areas (see Fig. 4.6) and their overlap tends to be high enough to be executed once the a force-closure grasp is found.

Figure 4.11 presents the time elapsed on the grasp planning until a feasible one is found that can be sent for execution. Results confirm those from Fig. 4.10 although it can be seen that for some objects (with ids 7 and 9), the MOOM planner needs

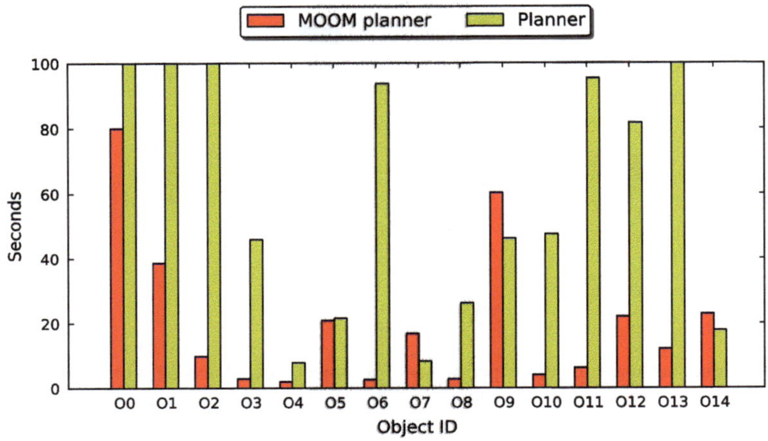

Fig. 4.11 Average planning time for each object until the first feasible grasp can be sent for execution for both (grasp plans taking more than 100 s are cut)

Table 4.2 Grasp success percentage for each planner. One trial per object

Planner	Objects grasped successfully	Success
MOOM		66.6 %
Default		46.7 %

more time to process less approach vectors. This is due to the fact that very bad grasp hypotheses can be filtered much faster than those with a higher goodness.

Finally, Table 4.2 presents grasp success rates for both planners regarding execution on the real platform. From the figure it can be seen that the planner using MOOM performs better than the standard planner—42.5 % of improvement—and in average needs about 10 times less time to find the first feasible grasp. To obtain a more representative evaluation, the objects are grasped again twice using the grasp planner with MOOM. Figure 4.12 presents the results of this evaluation and it can be seen the overall success rate is similar (2 % less). It is worth mentioning that even though success rates are rather low in comparison to state-of-the-art grasping techniques, we argue that there is a lot of room for improvement in our pipeline as: (1) the model of the objects do not match completely the real objects and there are no constraints on the objects or their configuration, (2) the first grasp found by the planner is sent for execution and (3) our starting point is a standard planner with very low success rates as seen in Table 4.2.

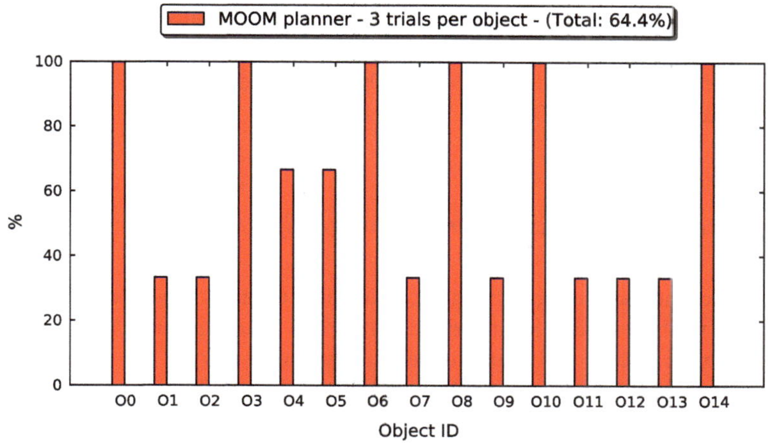

Fig. 4.12 Grasp success rate for MOOM grasp planner using 3 trials per object

4.3.5 Discussion

We have presented the Model-Object Overlap Metric (MOOM) to incorporate information obtained from the depth-sensor into the grasp planning process. We have shown how MOOM can be incorporated at different stages and in different fashions to conventional grasping pipelines to increase efficiency and robustness to errors from recognition.

To better account for mismatches between simulation and reality, reactive grasp controllers [24] could be activated to boost the grasp success rate. This and the implementation of MOOM in more advanced grasp planners are considered as possible future works.

4.4 Grasping Unknown Objects: Using Symmetry Assumptions

In this section, the scenario of picking up unknown objects from a table top is considered. To accomplish grasp and motion planning based on this information, the idea of filling in the gaps in the scene representation is followed through predicting the full shape of each object. We make use of the observation that many, especially man-made objects, possess one or more symmetries. Given this, the simulator can use an estimated complete world model under which it can plan actions or predict sensor measurements.

The work presented in this section is the result of a collaboration with Jeannette Bohg,[3] Javier Felip,[4] Matthew Johnson-Roberson,[5] and Xavi Gratal. A description of the use of the simulator is presented here but a detailed description of the complete approach can be found in [3].

4.4.1 Predicting Object Shape Through Symmetry

Psychological studies suggest that humans are able to predict the portions of a scene that are not visible to them through *controlled scene continuation* [25]. A very strong prior that exists in especially man-made objects is symmetry.

[26] showed that this symmetry can be detected in partial point clouds and then exploited for shape completion. *Planar reflection symmetry* is defined as the case in which each surface point P can be uniquely associated with a second surface point Q by reflection on the opposite side of a symmetry plane. Furthermore, in a household environment, objects are commonly placed such that one of their symmetry planes is perpendicular to the supporting plane. Exceptions exist such as grocery bags, dishwashers or drawers.

Given these observations, for our scenario we can make the assumption that objects commonly possess one or several planar symmetries of which one is usually positioned perpendicular to the table from which we are grasping. By making these simplifications, we can reduce the search space for the pose of this symmetry plane significantly.

We follow a generate-and-test scheme in which we create a number of hypotheses for the plane parameters and determine the plausibility of the resulting mirrored point cloud based on visibility constraints. We bootstrap the parameter search by initializing it with the major or minor eigenvector e_a or e_b of the projected point cloud, which usually yields a good first approximation. Further symmetry plane hypotheses are generated from this starting point by varying the orientation and position of the eigenvectors as outlined in Fig. 4.13.

Given these symmetry plane parameters, the mirrored point cloud is determined as follows and its visibility score is computed. We search for the global minimum in the space of all scores that corresponds to a reflected point cloud with the smallest amount of points that contradict the symmetry hypothesis. After the prediction of the backside of an object point cloud, we create a surface mesh approximation to support grasp planning and collision detection. In order to achieve that, the Poisson reconstruction proposed by [27] is used.

[3] Autonomous Motion Lab, MPI for Intelligent Systems, Germany.

[4] Robotic Intelligence Lab, Universitat Jaume I, Castellon, Spain.

[5] Computer Vision and Active Perception lab, KTH, Sweden.

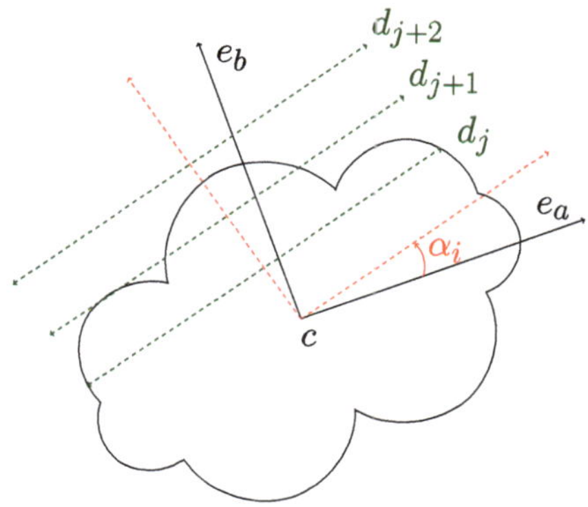

Fig. 4.13 A set of hypotheses for the position and orientation of the symmetry plane. e_a and e_b denote the eigenvectors of the projected point cloud. c is its center of mass. α_i denotes one of the variations of line orientation along which the best pose of the symmetry plane is searched. The (*green*) lines at positions d_j to d_{j+2} are three further candidates with orientation α_i

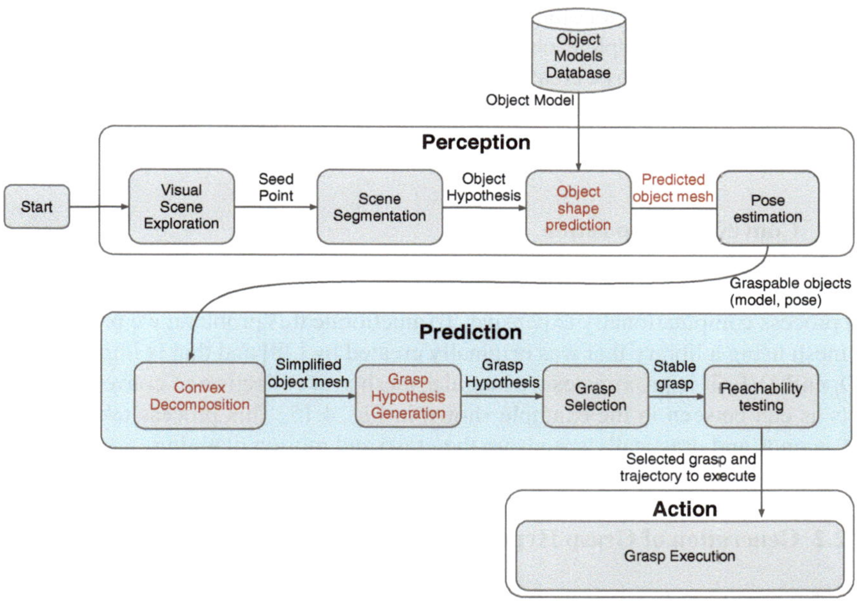

Fig. 4.14 Grasp pipeline for grasping unknown objects using simetry assumptions. Alterations of the referenced pipeline (Fig. 4.4) are shown in *red*

4.4.2 Grasping Pipeline

To emphasize the generality of the proposed approach, it was evaluated and demonstrate it on two robotic platforms being either used at the Royal Institute of Technology (see Sect. 3.3.5) or at the Universitat Jaume I (see Sect. 3.3.5). For the scenario of grasping unknown objects from a table top, an overview of the grasp cycle is shown in Fig. 4.14.

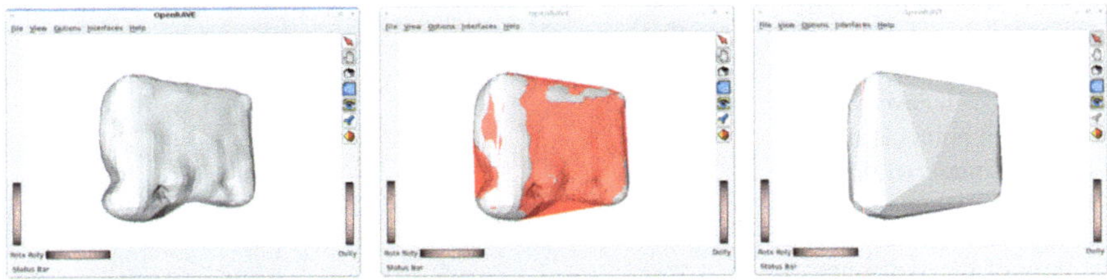

Fig. 4.15 Example of convex hull decomposition of the generated triangular mesh

As a step prior to the shape analysis of an object and its grasping, the robot needs to explore the scene and segregate potential objects from the background, returning a segmented point cloud. For collision detection that is necessary for motion planning, the simulator needs complete object models. In the case of unknown objects, an approximate model has to be created on-line. Given the segmented point cloud which only represents the visible part of an object, its backside can be predicted with the method described in the previous section.

4.4.2.1 Convex Decomposition

The obtained object mesh has thousands of vertices which makes the collision detection process computationally expensive. To ameliorate this problem, we pre-process the mesh using a library that was originally created by [28] and that is implemented in OpenRAVE. It approximates a triangular mesh as a collection of convex components as can be seen in the example shown in Fig. 4.15. This process takes only a few seconds and drastically speeds up the grasp and motion planning.

4.4.2.2 Generation of Grasp Hypothesis

With the complete object shape as an input, a set of grasp candidates to be tested is generated and evaluated in simulation. Using the default algorithm of OpenRAVE to generate grasp candidates, the time to simulate all the corresponding grasps can vary from few minutes to more than an hour. These execution times are acceptable for objects that are known beforehand because the set of grasp candidates can be generated off-line. When the objects are unknown, this process has to be executed on-line and long waiting times are not desirable.

For this reason, we use two methods to reduce the number of approach vectors. The first one, applied on the KTH platform, computes two grasp points as in [29]. To grasp the object at these points, there are an infinite number of approach vectors on a circle with the vector between the two grasping points as its normal. We sample a given number, typically between 5 and 10 of these between 0° and 180°. Figure 4.16c shows an example of the detected grasping points along with the generated approach

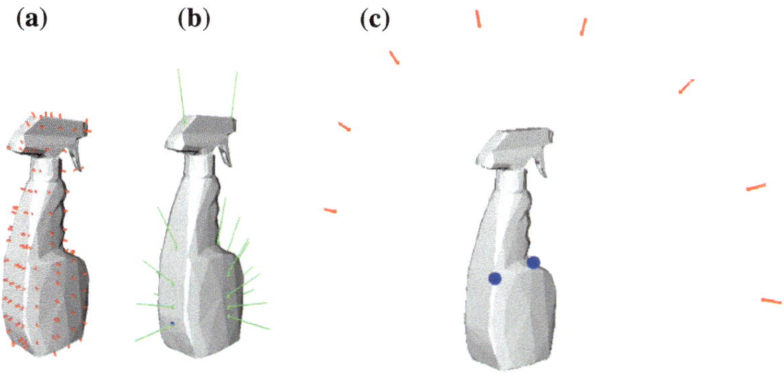

Fig. 4.16 Example of the approach vectors generated for a spray bottle by (**a**) the OpenRAVE grasper plugin, (**b**) the UJI proposed algorithm using the object's centroid and (**c**) the KTH proposed algorithm using the grasp points in *blue*

vectors. The second method, used at UJI, calculates the approach vectors in a similar way only that the center of the circle is aligned with the object's centroid and its major eigenvector e_a. Another circle, perpendicular to the first one, is added in order to compensate the possible loss of vector quality due to the lack of grasp points. Figure 4.16b shows an example.

Having the list of approach vectors reduced, the other parameters were also adjusted for our purposes. As a hand pre-shape, we defined a pinch grasp for each hand. The approach distance is varied between 0 and 20 cm. Finally, the roll is chosen dependent on the two grasping points (such that the fingers are aligned with them) or on the orientation of the circle on which the selected approach vector is defined. Using these parameters, we were able to reduce the amount of time taken to generate and save the set of grasp candidates to less than a minute.

4.4.2.3 Grasp Selection

The next step consists of selecting a stable grasp from the set of grasp candidates that can be executed with the current robot configuration without colliding with obstacles. This process is the same as for known objects described in Sect. 4.2.2.

After a suitable grasp and arm trajectory has been selected through simulation, it is executed by the robot in an open loop procedure. If the robot successfully grabs the object and moves it to the destination, the stable grasp is returned for execution with the real robot. Otherwise, the next stable grasp from the set is tried.

4.4.3 Experiments

The database we used for evaluating the point cloud mirroring method is shown in Fig. 4.17. For each of the objects in this database, with the exception of the toy tiger and rubber duck, we have laser scan ground truth.[6] The test data was captured with the KTH vision system and contains 12 different household or toy objects.

From all point clouds, we reconstructed the complete meshes based on the method previously described. We used the two different values of 5 and 7 as the octree depth parameter of the Poisson surface reconstruction. By limiting this parameter, we enable mesh reconstruction in near real-time. With a tree depth of 5, the meshes are more coarse and blob-like but less sensitive to noise in the point cloud and normal estimation. With a depth of 7, the reconstructed surface is closer to the original point set. However, outliers strongly affect the mesh shape and it is more sensitive to noise.

To obtain the ground truth pose for each item in the database, we applied the technique proposed in [30]. It allows to register the laser scan object meshes to the incomplete point clouds.

As a baseline, we reconstructed a mesh without mirroring. To do this, we applied a Delaunay triangulation[7] to the projection of a uniformly sampled subset of 500 points from the original point cloud. Spurious edges were filtered based on their length in 2D and 3D. Furthermore, we extracted the outer contour edges of this triangulation and spanned triangles between them which produced a watertight mesh. Figure 4.18 shows the result of this Delaunay based mesh reconstruction for the toy tiger.

To assess the deviation of the Delaunay meshes and the mirrored meshes from the ground truth we used MeshDev [31]. As a metric we evaluated geometric deviation, i.e., the distance between each point on the reference mesh to the nearest neighbor on the other mesh. We applied the uniform sampling of the surface of the reference mesh as proposed in [31] to calculate this deviation.

Figure 4.19 shows the mean and variance of the mesh deviation between the ground truth mesh and the reconstructed meshes for all object orientations over all objects. We can state that the mirrored point clouds are on average always deviating less from the ground truth than the Delaunay based meshes. The average deviation for the mirrored meshes over all orientations amounts to 7 mm.

We demonstrated the approach proposed in this paper on two robotic platforms.[8] Figure 4.20 shows snapshots of the grasp execution in simulation on the predicted objects and with the real robots on the real objects.

At KTH, several objects were placed on the table emphasizing the benefit of object shape prediction for motion planning. Furthermore, it showed that the prediction mechanism can deal with some occlusions. This is due to the enforced visibility constraints. One of the main differences between the runs at UJI and KTH was the resolution of the point clouds that is due to the use of camera systems with different

---

[6] The ground truth object models were obtained from http://i61p109.ira.uka.de/ObjectModelsWebUI/.

[7] http://opencv.willowgarage.com

[8] A video of the experiment can be found at http://youtu.be/jskDy2IfQr4.

Fig. 4.17 The 12 Objects in the Database in varying poses yielding 16 data sets. Objects are shown in their 0° position, i.e, with their longest dimension parallel to the image plane. Ground truth meshes are existing for all objects except the toy tiger and the rubber duck

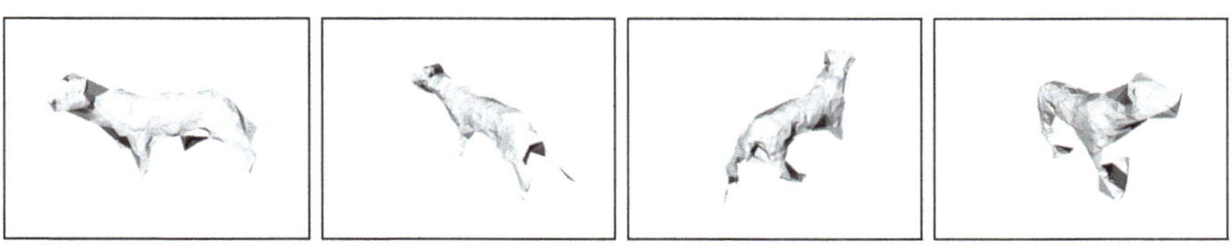

Fig. 4.18 Delaunay based Meshes of toy tiger in the following orientations: $0°, 45°, 135°, 225°$

focal lengths. While the KTH point clouds usually consist of 40,000 points, UJI point clouds contained around 3,000 points. However, a suitable mesh could still be generated with the advantage of a lower runtime. When running the whole generate and test procedure (with $n = 6$ and $m = 5$ yielding 35 hypotheses) on a single core of an I7 CPU with 2.8 GHz, we achieved the following run-times: 16.46 s for a point cloud with 39,416 points and 0.31 s for a point cloud with 2,100 points. Please note that we have not exploited the possibility to parallelize this process yet. These runtimes also show that downsampling the point clouds before mirroring can speed up the shape completion without a big loss of precision. To investigate these optimizations is considered as future work.

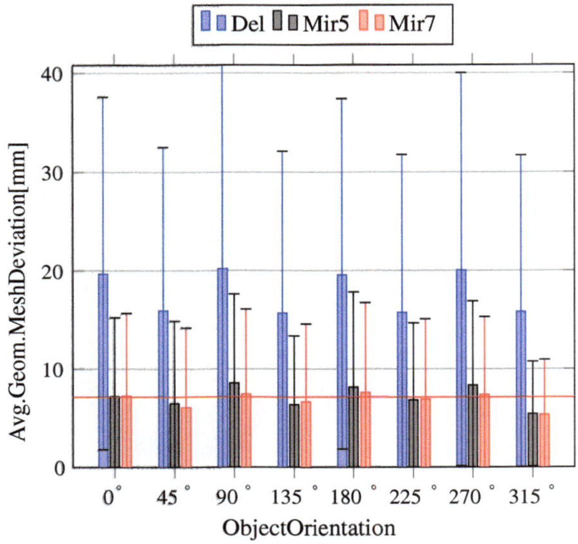

Fig. 4.19 Evaluation of the deviation between the Ground Truth Mesh and (1) Mesh based on 3D Point Cloud only (*Del*), (2) Mesh based on mirroring and Poisson Surface Reconstruction with Tree Depth 5 (*Mir5*) or (3) with Tree Depth 7 (*Mir7*)

4.4.4 Discussion

The technique proposed in this section is a first step towards bridging the gap between simulation and the real world. The experimental results show that even if the object model is not an exact representation of reality, it is close enough to enable the simulator to try different grasp alternatives and select an appropriate one.

However, in the presence of strong occlusions the proposed method will most likely fail to estimate the correct object surface. Among others, this is due to using the eigenvectors of the point cloud projection as an initialization for the underlying symmetry plane. This choice has been shown on the dataset to yield reasonable estimates for the orientation of the symmetry plane. However, if the point cloud represents only a minor part of the real object, then the eigenvector of this part will not be close to the optimal symmetry plane of the real object. Also wrong segmentations can lead to erroneous results. The dataset on which we quantitatively evaluated the proposed method contains only single objects.

Another possibility is to keep the predicted object shape but explicitly take the uncertainty about the reconstructed surface into account for grasp and motion planning. This could be done by for example avoiding placements of the hand in which fingertip positions coincide with points of low plausibility. This approach is shown in the following section.

(a)

(b)

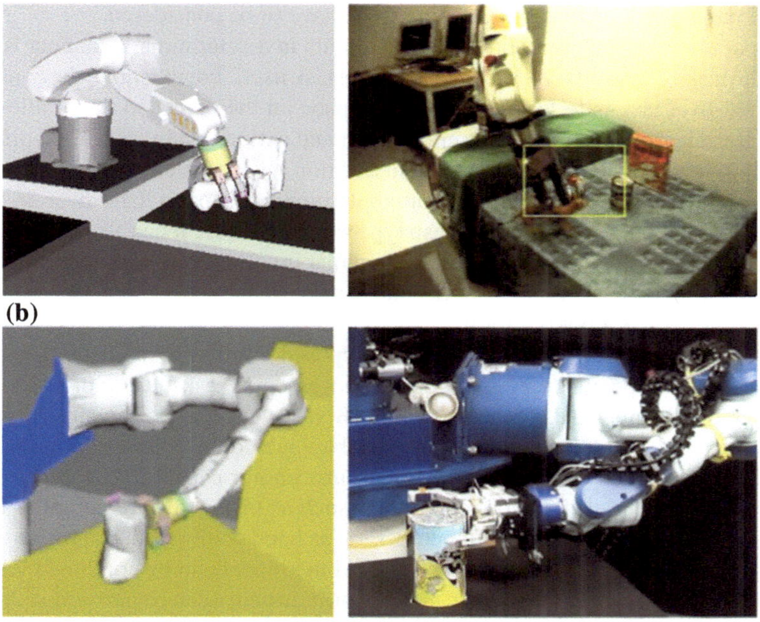

Fig. 4.20 Examples of the grasp performed by the simulated robot and the real one. **a** KTH platform. **b** UJI platform

4.5 Grasping Familiar Objects: Using Task Constraints

In the previous approach, the object was considered to be unknown to the robot and it was reconstructed based on symmetry assumptions. In this section, we start with the assumption that the object is a familiar object, which means that it is geometrically similar to a known model. The objective of this approach is to achieve autonomous grasping of objects according to their category and a given task.

Recent advances in the field of object segmentation and categorization as well as task-based grasp inference have been leveraged by integrating them into one pipeline. This allows us to transfer task-specific grasp experience between objects of the same category.

Robot grasping needs to be goal-directed. An autonomous robot needs to plan and execute a grasp that affords next step of manipulation. For example, a robot shall not grasp from the top of a mug when the task is to pour out the water, and shall not grasp the handle of a knife if the task is to hand-over it. These semantic task requirements constrains, in a complex manner, the planning and execution of each grasp. The robot has to consider not only the hand pose, finger configuration, but also the pose of the object, and its physical properties that determine if the object affords this task. Also, to firmly grasp an object for further manipulation, important control parameters such as the grasping force have to be considered.

To parametrize such semantic task constraints in a deterministic manner is hard. First, the task constraints can vary a lot with the task itself. For example, the constraint of a hand-over task is to leave enough free-space on the object so that it allows re-grasp. It is clearly described by a set of object and action variables that are different from those that define the pouring task. In addition, this task description may also be hand-specific. For example, human can apply power grasps to hand-over an apple, but a robot may fail with the same grasp type simply because it has a larger hand. In a nutshell, the grasping task space is a domain that involve many variables with high-dimensionality and the good grasps depend heavily on the sensorimotor embodiment of every individual robot platform.

Some recent work in the area [32, 33] proposed a fully probabilistic framework for embodiment-specific representation of robot grasping tasks. It exploited strengths of the graphical model, Bayesian networks (BNs) to encode the semantic task constraints on a multitude of task relevant variables including object, action and constraint features. Such a system encode the sensorimotor capability for each robotic platform. The high-level symbolic variables such as task goals are naturally linked to the low-level sensorimotor parameters such as hand configuration through conditional probabilistic distributions. Once trained the model can be combined with the probabilistic decision making where grasp plan and control can be performed through inference even with noisy and partial observations. The framework has been shown to successfully make goal-directed grasp planning in simulation environments.

Very few grasping systems have approached the problem of transferring grasp experience between objects of the same category. Marton et al. [34] present and demonstrate an approach similar to ours in that an object categorization system is used in a robotic grasping framework. Furthermore, it combines 2D and 3D descriptors. However, the category of an object is not used to infer a suitable grasp. Instead, a 3D descriptor helps to narrow down the choice of categories to those of similar shape, and then a 2D descriptor is applied to look up a specific object instance. One of the grasp hypothesis associated to that object instance is then executed.

In this section, we demonstrate how integration of 2D and 3D cues for object categorization can facilitate robot grasping. Specifically, we do not require exact models of the object to be grasped. A model of the same category annotated with grasp hypotheses is sufficient. Furthermore, we show how the detected category helps to infer a grasp that is task-specific. The effectiveness of the approach is demonstrated on the humanoid robot ARMAR-IIIa (Sect. 3.3.5) and on the Tombatossals robot (Sect. 3.3.5).

This work is the result of my collaboration with several researchers: Dan Song,[9] Jeannette Bohg, Marianna Madry, Javier Felip,[10] Kai Welke,[11] Martin Do, Markus

[9] Computer Vision and Active Perception lab, KTH, SE.

[10] Robotic Intelligence Laboratory, UJI, ES.

[11] Humanoids and Intelligence Systems Lab, Institute for Anthropomatics, KIT, DE.

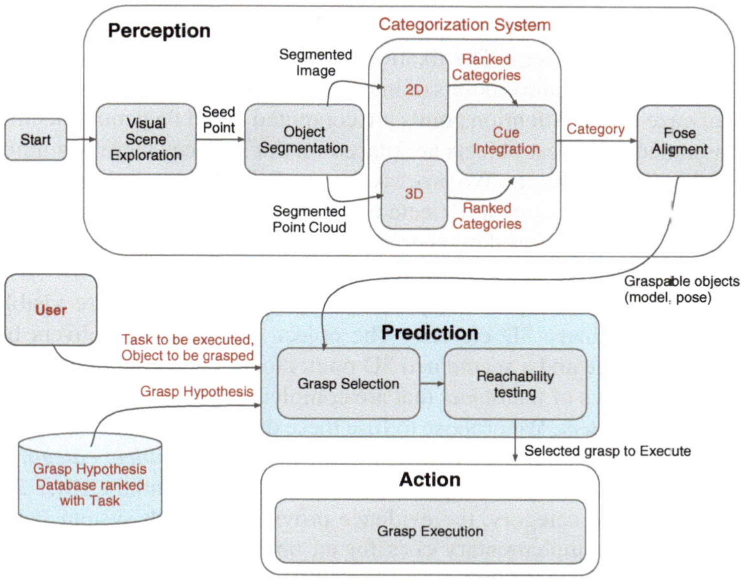

Fig. 4.21 Grasping pipeline for graping familiar objects using task constraints. Alterations of the referenced pipeline (Fig. 4.4) are shown in *red*

Przybylski, Walter Wohlkinger[12] and Aitor Aldóma. A description focusing on the use of the simulator is presented here but a detailed description of the complete approach can be found in [4].

4.5.1 Grasping Pipeline

An overview of the proposed grasping pipeline is shown in Fig. 4.21. It is subdivided into three major building blocks. Given visual input from stereo cameras, the perception block is responsible for detecting an object and estimating its category and pose. The prediction block takes in the object pose and category and infers a ranked list of grasps according to a specific task. The best grasp is then executed on the robotic platform.

4.5.1.1 Perception

Attention. The success of visual processing such as classification rate and reliability of pose alignment strongly depends on the quality of the visual sensor data provided

[12] Automation and Control Institute, TUW, AT.

as input. Consequently, providing as detailed views of the objects as possible is beneficial for all processing steps. The fixation of the objects involves mechanisms of attention allowing to determine points of interest in the scene and mechanisms for the execution of gaze shifts. Attention points are computed based on geometric information. It is assumed that most objects are placed on flat surfaces thereby simplifying the detection of interest points. We process the resulting 3D data using plane fitting similar to [12]. After removing the detected support surface, the remaining 3D points are clustered in an unsupervised manner using the growing neural gas method [35]. Each cluster center serves as a point of interest.

Segmentation. Once the previously detected points of interest are visible, we can refine the preliminary 3D clusters. The object segmentation delivers both, a segmented RGB image and a segmented 3D point cloud. Each kind of data encodes different characteristics of the object that are complementary (see Figs. 4.22a, b).

Categorization System. We propose to fuse these different cues to achieve a more robust object categorization. Specifically, we run two distinct *object categorization systems* (OCS) in parallel of which one is processing 2D cues and the other 3D cues only. To obtain a final category, the evidence provided by each system is merged (see Fig. 4.22c). As complementary cues for an object category, we have defined appearance (SIFT [36]), color (opponentSIFT [37]) and 3D shape [17].

Pose Alignment. Whether a grasp is suitable for an object of a specific category and task, is highly dependent on the relative alignment between hand and object. Therefore, we need to estimate the exact pose of the object in the scene. From the previous categorization step, we are given a set of object prototype models of a specific category that are most similar to the current object of interest. To determine the best matching model from this set, we need to align each of them with the segmented point cloud and compare the values of a similarity metric.

4.5.1.2 Prediction

At this point in the processing pipeline, we have the following information available about an object in the scene: (1) its category, (2) the most similar object model from a database and its estimated pose and (3) a specific task. Given this, our goal is to infer a ranked list of grasps.

We approach this problem, by *off-line* generating a set of task-ranked grasp hypotheses. This involves (1) the generation the grasp hypotheses and (2) their ranking according to the task and the object's category. In the *on-line* process, the most similar object model and the given task serve as a look-up in this database to retrieve the highest-ranked grasp. This is visualized in Fig. 4.21 in the *Prediction* block. These building blocks are described in more detail.

Grasp Hypothesis Generation. This process is performed off-line using the grasp planning method proposed by Przybylski et al. [38] with OpenRAVE and is based on the medial axis transform (MAT) [39]. The MAT can represent arbitrary three-dimensional shapes. It is constructed by inscribing spheres of maximum diameter into an object's shape, where each of theses spheres have to touch the object's surface

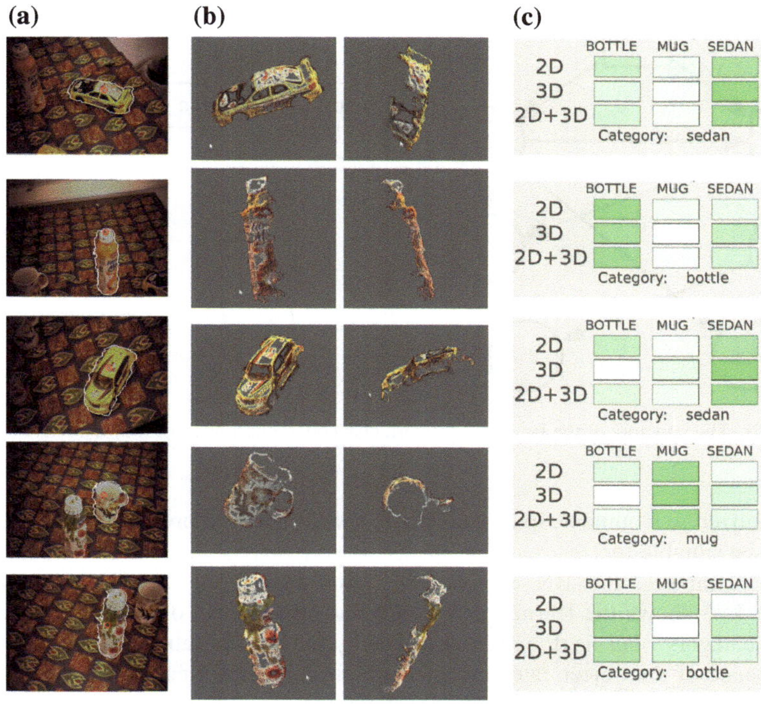

Fig. 4.22 Intermediate results of the categorization pipeline. The more saturated the green, the more confident is the categorization. **a** Segmented object and seed point (*red*). **b** Reconstructed point clouds. **c** Categorization results

at two or more different points. The object is then described as a set of spheres, where each sphere has a position, radius and an angle between its centroid and the closest object boundary point as parameters [40]. For the actual grasp planning process, we sort the inscribed spheres into a grid structure with respect to their Cartesian coordinates. Candidate grasps for a sphere in this grid are generated by estimating the symmetry properties of the sphere centers in the query sphere's vicinity. These symmetry properties are then used to choose approach point, approach direction and hand orientation such that the fingers can wrap around the object. Each candidate is tested for stability using the common ϵ-measure for force-closure [41].

Task Constraint Model and Task Ranking. We model the conceptual task requirements for a given hand through conditional dependencies between the task T and a set of variables including object features O, grasp action parameters A and constraint features C. This is captured in a Bayesian Network (BN). As described in more detail in [32, 33], both the parameters and the structure of this BN are learned. The necessary training data for this process is generated from a synthetic database of objects on which grasps have been generated. Each of these object-grasp pairs

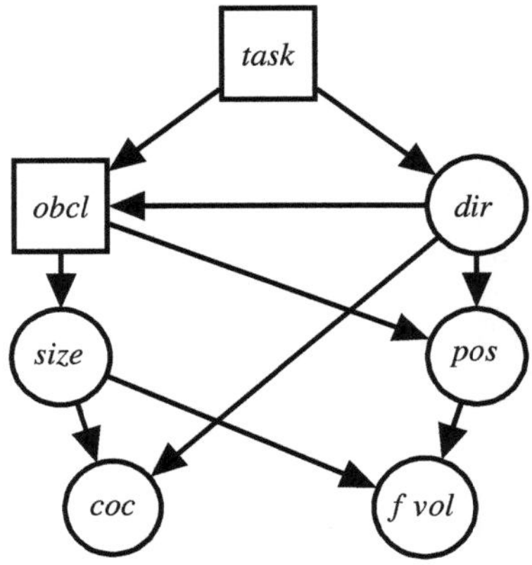

Groups	Name	Dim	Description
T	*task*	-	Task Identifier
O	*obcl*	-	Object Category
	size	3	Object Size
A	*dir*	4	Hand Orientation
	pos	3	Grasp Position
C	*coc*	3	Center of Contacts
	fvol	1	Free Volume

Fig. 4.23 The structure of the Bayesian network task constraint model for the ARMAR hand

is visualized to a human expert who labels it with one or more tasks that this grasp would be suitable for.

After training, the BN encodes the joint distribution of all the variables. Figure 4.23 shows the learned task constraint model. One of the object features, object category *obcl*, is directly influenced by the task variable. This indicates the importance of the object category information in determining its functional affordance of grasping tasks, hence reinforces the importance of the object categorization in the "perception" block.

Given this BN, we can then infer the conditional distribution of one variable given observations of all or a subset of other variables. For example, we can infer $P(obcl|task)$ to decide on which object category the given task can be best performed. We can also infer $P(A|task, obcl)$ to provide grasp ranking given a task and the category of the current object of interest.

Grasp Selection based on Object Category and Task. The result of the off-line process provides the robot with databases of grasp hypotheses for each object annotated with a task-probability. Given this, we can select the best grasp to be executed on-line.

The perception part of the pipeline outputs the perceived objects in the scene with their calculated poses in the environment and they are loaded into OpenRAVE. A task is then specified by the user. The grasp hypotheses stored for that object are ordered by their probability for the specific task. The best ranked hypotheses are then tested to ensure that they are reachable under the current robot configuration and that the robot trajectory is collision free. Once the best rated, reachable grasp hypothesis has been determined, it is executed by the humanoid platform. Figure 4.24 shows an example scene on which we illustrate the different steps of the proposed pipeline.

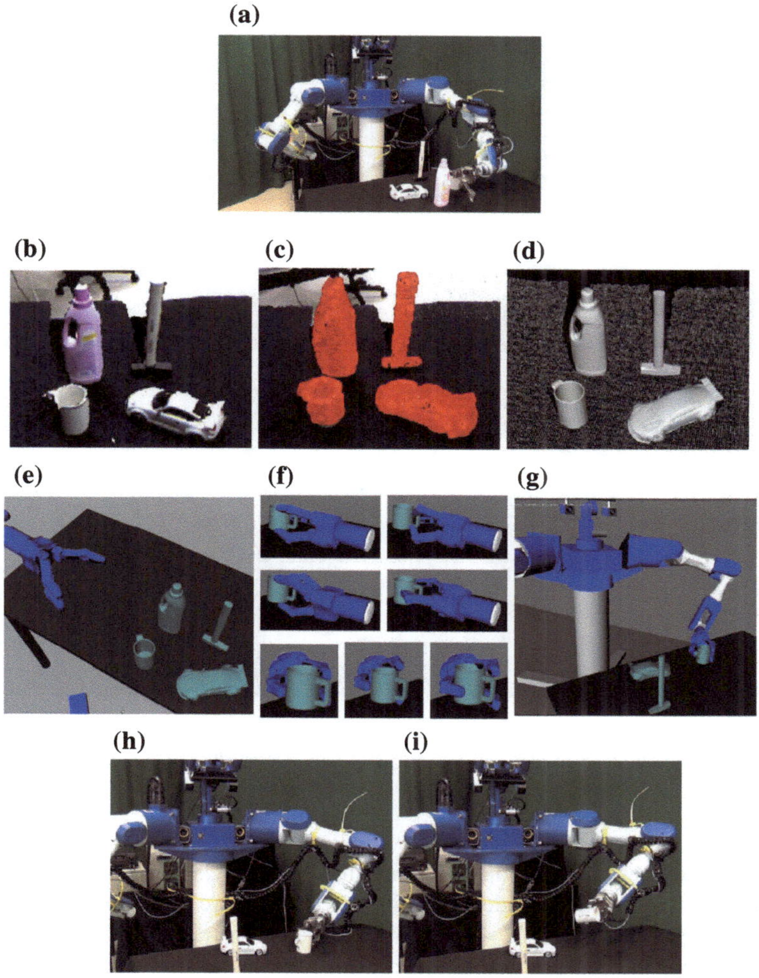

Fig. 4.24 Example of graping pipeline for a mug when the task selected is *pouring*, using the humanoid Tomabatossals. **a** Example scene. **b** Visual exploration. **c** Object segmentation. **d** Object recognition and alignment. **e** Categorized objects loaded in the simulator. **f** Grasp selection. **g** Reachability test for the selected grasp. **h** Executing grasping. **i** Executing pouring task

4.5.2 Experiments

Our experimental setup features a scene which contains several object instances of different categories placed on a sideboard. We consider objects of four different categories: cars, bottles, mugs and hammers. The 2D OCS system was trained on two object instances per category. For each object, RGB images were collected from eight

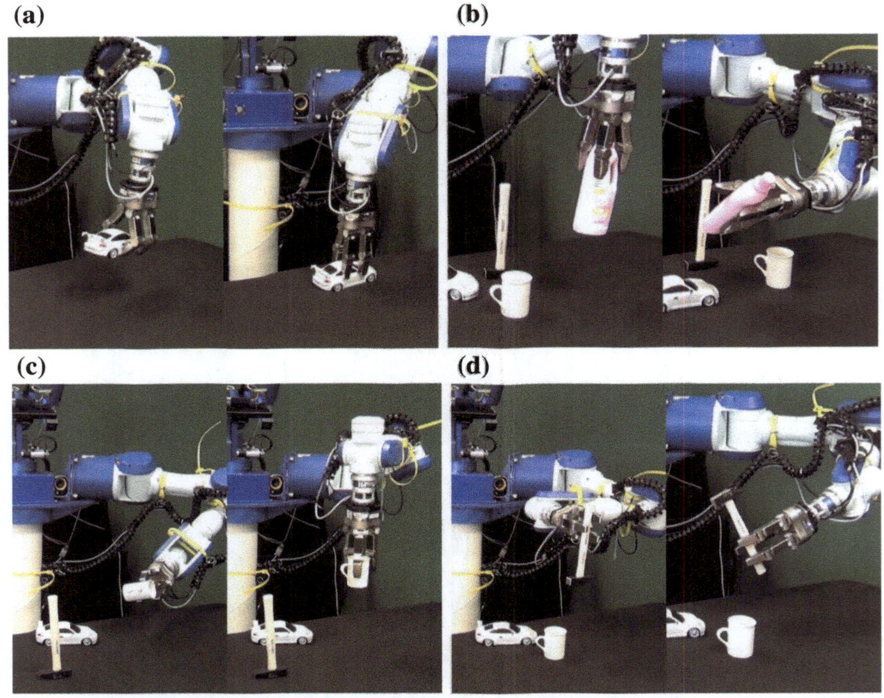

Fig. 4.25 Example grasp executions of grasps given different tasks using Tombatossals. a Hand-over vesus playing. b Hand-over versus pouring. c Pouring versus hand-over. d Hand-over versus tool-use

equidistant views around the object. For training the 3D OCS, seven car, eight bottle, ten mug and five hammer models each in two scales were used. Eighty views per model were generated. In [17, 42], each OCS is demonstrated for a much broader class of categories. The restriction to only four classes in this work is due to the extensive task-labeling process as well as to restrictions on the size and weight of the objects to be grasped.

We are considering five different tasks: hand-over, pouring, dish-washing, playing and tool-use labeled on 1956 stable grasp hypotheses for the 60 object models. Since some hypotheses are good for multiple tasks, we have 2,166 training examples in total.

We directly compared the execution of grasps on the same object in the scene given different tasks using the two robot platforms: Tombatossals' examples are shown in Fig. 4.25 and ARMAR-IIIa in Fig. 4.26.

It can be observed that grasps for the task *hand-over* are usually top grasps that leave a major part of the object uncovered. The grasps for *playing*, *pouring*, *dish-washing* and *tool-use* are as expected oriented towards the functionality of the object itself in relation to the task. Using ARMAR-IIIa for the dish-washing task, the best

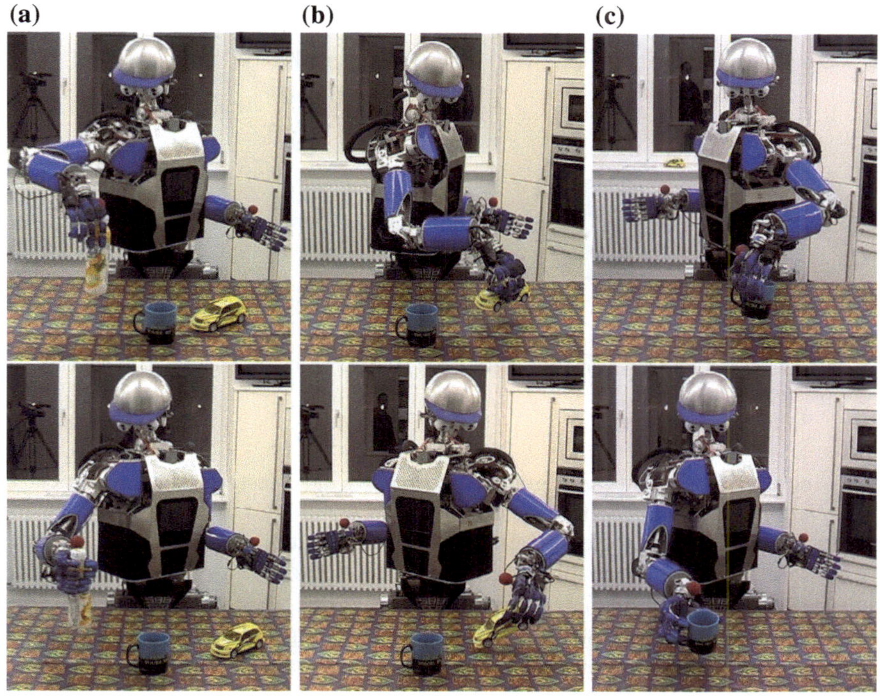

Fig. 4.26 Example grasp executions of grasps given different tasks using ARMAR-IIIa. **a** Hand-over versus pouring. **b** Hand-over versus playing. **c** Pouring versus hand-over

ranked grasp on the mug is visualized in Fig. 4.27 (*right*). This grasp was however not reachable in this scene and therefore the next best grasp for this combination of object and task has been selected. It is very similar to pouring from this mug as visualized in Fig. 4.27 (*left*).

4.5.3 Discussion

In this section, we presented a grasping pipeline that allows autonomous robot grasping according to an object's category and a given task. Several state-of-the-art modules performing scene exploration through gaze shifts, segmentation, object categorization and task-based grasp selection were integrated. We showed how this allows the robot to transfer task-specific grasp experience between objects of the same category. The effectiveness of the pipeline was demonstrated on two humanoid robots.

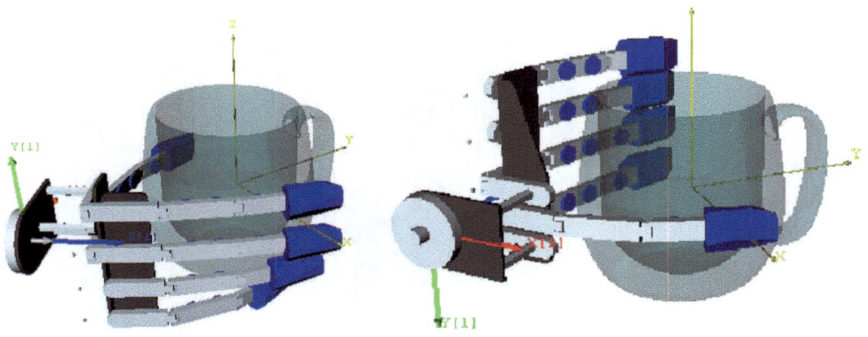

Fig. 4.27 Comparison of best ranked grasp for a mug according to the task *pouring* (*left*) or *dish-washing* (*right*)

To increase the robustness of grasp execution, we have designed and implemented an *overcomplete* pipeline where the task of different modules overlap. This holds for attention, segmentation, categorization and pose estimation.

However, information in this pipeline only flows into one direction (from left to right) without any intermediate failure detection. If this was the case, repeated execution of perceptual processes could be requested to improve the input to a module. Furthermore, this repetition could be based on more information that was already obtained in a later stage of the pipeline.

Another potential improvement of the robustness of the pipeline could be achieved by not only executing the reaching motion in a closed loop manner but also the grasp itself. From the perception block of the pipeline, we know the geometry of the object quite well. This would allow us to adopt an approach similar to [43] for on-line comparison of actual and expected tactile sensor readings and adaptation of the grasp if necessary.

4.6 Dynamic Grasping Simulation

In this section, the challenge of developing the full dynamics simulation of a complete robot is addressed, including the dynamics of the bodies, the actuation of the motors and the simulation of the sensor readings.

In the context of grasping and dexterous grasping, simulation has been mostly limited to replicating the kinematics and not the dynamics of the robot manipulators. Dynamics simulation, which takes into account masses, forces, inertias, static and dynamic frictions, and even elasticities and deformations, is a very challenging problem, specially when it comes to considering the interactions between contacting bodies. A large number of parameters, which are difficult to determine, affect the dynamics behaviour of the involved parts.

A complete dynamic simulation of a robot can be of great benefit for robotics research. First, the simulation could replace the real hardware to the extent that it reproduces the actual physical behaviour, which is of special importance in the context of robot manipulation. Second, it can be used as an accurate prediction engine that can help us to understand the effects of actions and be the base for developmental learning. Additionally, if simulation accurately reproduces the real sensor and actuator feedback, robots could automatically learn from low-level sensor inputs without the erosion of real hardware.

There have been some attempts to create a simulation of robot dynamics [44] described a method to optimize a set of simulation parameters using evolutionary algorithms in the context of mobile robotics [45] studied how accurately the pushing of flat objects across a table can be predicted by a physics engine adapting some of its parameters to enhance the simulation [46] simulated planar grasping actions using experiments to calibrate the system and then evaluating the results using a hand with one degree of freedom. However, these simulations have been performed using very simple and controlled experiments in order to reduce the complexity of dynamics simulation.

The purpose of this work was to develop a complete framework using existing tools to assess the suitability of simulation to serve as a surrogate for real robotic grasping system in order to provide full dynamic simulation of manipulation tasks in virtual scenarios.

Most importantly, three of the most representative manipulation tasks are selected and implemented to compare the real behaviour with the simulated one. There has been little work published about dynamics simulation of grasping, and what has been done, performed experiments with very tight constraints. For this reason, the proposed experiments are mainly focused on evaluating the simulation behaviour of performing realistic robot manipulation tasks. These tasks are: object grasping and lifting, in-hand manipulation and object pushing/sliding over a surface. The sensor and actuator's feedback has been recorded for both real and simulated systems, comparing and analysing the results. Finally, we report the degree of success in each of these experiments and the accuracy of the results obtained. This work has been the result of my collaboration with Javier Felip[13] and Higinio Martí.

4.6.1 Implementation

The real robotic platform used to validate the dynamics simulation is the UJI humanoid torso Tombatossals (presented in Sect. 3.3.5).

[13] Robotic Intelligence Lab, Universitat Jaume I, Castellon, Spain.

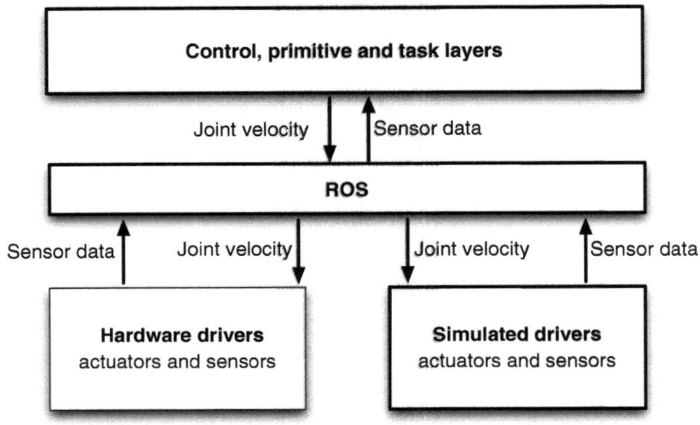

Fig. 4.28 Control architecture used to test the dynamics simulation

4.6.1.1 System Architecture

The architecture that controls the system is separated in low and high level (Fig. 4.28). The low level receives joint velocities, sends them to the motors and provides the sensor data to the high level. It has been implemented in C++ and makes use of the Robot Operating System (ROS) for inter-module communications. The simulated robot implements its own low level system using exactly the same inputs and outputs as the real robot low level system. Thus the upper layers can transparently run any controller without knowing if the robot they are controlling is real or not. To validate the simulation, several tasks were defined and exactly the same controllers were executed on both the real and simulated systems.

4.6.1.2 Simulation

The simulated representation of Tombatossals was created to work with OpenRAVE. The plugins and sensors created for the dynamics simulation are available inside the robot manipulation toolkit OpenGRASP [47]. The physics engine used is ODE through the plugin available in OpenRAVE. However, the collision forces between the tactile sensors and the objects were calculated using the tactile sensor plugin from OpenGRASP. This calculated forces were then applied to the colliding bodies using ODE.

The geometrical representation of the robot was obtained from different sources and modified to adjust it to the values of the real robot. The torso and the head CAD models were created according to the measurements of the real robot. The models of the Mitsubishi PA10 arms and the Barrett hand were taken from the OpenRAVE robot model database and modified to fit the exact model. Finally, the Schunk hand CAD model was provided by the manufacturer.

Sensors

The tactile sensors were simulated using the developed tactile sensor plugin (Sect. 3.4) for OpenRAVE available in OpenGRASP. The model of the tactile sensors considers soft contacts and a full friction description including stick-slip phenomena. The sensor model consists of a surface contact patch described by the mesh of the contact elements. Therefore, meshes with the appropriate geometry for each of the robot tactile sensors were created. The parameters needed to adjust the model of the tactile sensors were the static and dynamics friction, and the stiffness K. The cameras were modelled using the camera sensors provided in OpenRAVE with the same intrinsic parameters and position as the real cameras.

Actuators

The angular motors, available in the robot arm and hand joints, have been simulated with the ODE controller provided by OpenRAVE. Each simulated servo-motor is parameterized by the maximum speed, the maximum acceleration and the maximum torque that the motor can apply. To control each motor, an interface to the ODE velocity controller plugin has been developed to enable the simulated robot to receive joint velocity commands from the controller through ROS, allowing the same control messages to be used by both robots.

4.6.2 Experimental Setup

Three different manipulation tasks were used to evaluate the similarity between the simulator and the real world. These tasks were selected to have different types of interaction between the robotic hand and the manipulated object: grasping, in-hand manipulation and sliding. For each manipulation task, a different object is used. The experimental environment consists of the Tombatossals robot in front of a table and next to a wall on its left. For these experiments only the left arm of the robot (Fig. 4.29) was used.

The evaluation of the simulator is performed continuously alongside the execution of each task. Data from each task is gathered at a constant rate. Each sample from the real execution is compared with its corresponding from the simulated one. The grasp task is executed five times while the sliding and the in-hand manipulation are executed once.

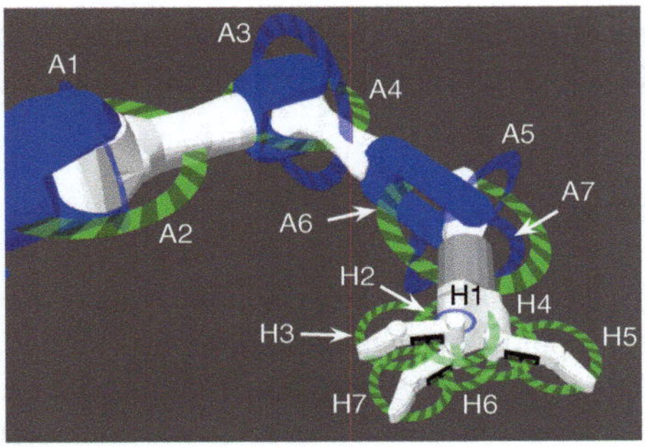

Fig. 4.29 Arm (*A*) and hand (*H*) joints of the Tombatossals right arm

4.6.2.1 Manipulation Tasks

Grasp

This task consists of grasping a box (116 × 103 × 218 mm, 143 g), lifting it 10 cm and placing it down again. The movement and transport primitives are basically devoted to move the arm to the desired position. Meanwhile the grasp controller closes the hand until a certain force is felt with the tactile sensors. The grasp is executed from five different positions starting from a top grasp and rotating the hand around the object towards a lateral grasp (see Fig. 4.30a).

Slide

This task consists of sliding an empty pizza box (320 × 370 × 55 mm, 263 g) from the initial position to the target position 35 cm away as shown in Fig. 4.30b. The slide controller looks for a first contact with the tactile sensors. While the contact is detected the arm moves towards the target position. If there is too much force detected the arm moves up and if the detected contacts disappear the arm moves downwards until they are detected again. This behaviour continues until the hand reaches the target position.

In-hand Manipulation

This task consists of swinging a wooden stick (60 × 530 × 16 mm, 370 g) that is grasped by the robot. The starting setup is the robot holding the stick as shown in Fig. 4.30c.

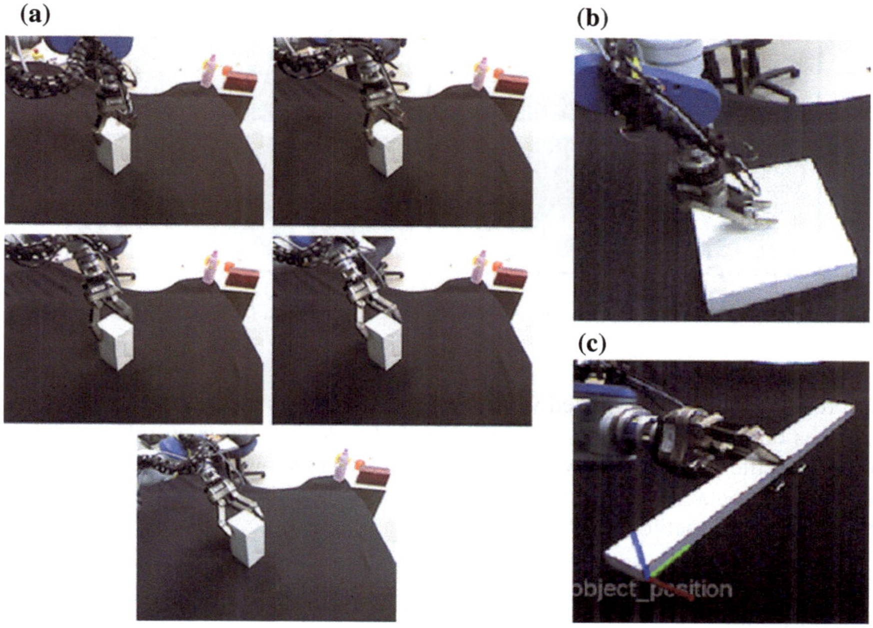

Fig. 4.30 Manipulation tasks: starting position. **a** Grasping. **b** Object sliding. **c** In-hand manipulation

4.6.2.2 Metric Definitions

A set of metrics to evaluate the similarity between the simulator and reality has been defined for each component of the simulated robot. The compared values are: arm joint values, hand joint values, manipulated object position and tactile sensor readings.

Joint Values

The joint values of the arm and the hand are logged to validate that the simulated actuators behave like the real ones. Although the similarity may not be 100 % accurate, a similar behaviour during all the manipulation tasks would be enough to conclude that the simulator can be used to replace the reality in at least such conditions.

Tactile Sensor Readings

Comparing tactile sensor readings would not only provide the precision of the simulated sensors but the difference between the instants of detection of contacts and the

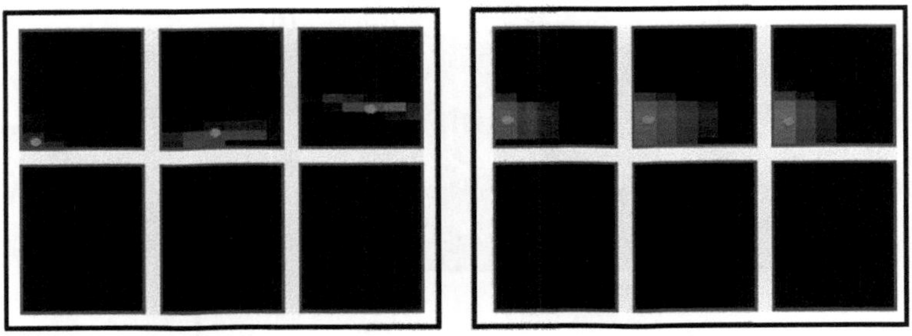

Fig. 4.31 Tactile sensor data with centroids during the grasping task. *Left*: Real. *Right*: Simulated

difference between the detected values. Figure 4.31 shows an example of the tactile readings during the grasping task, the figure also depicts the centroid as red dots on each tactile patch. As the variations in intensity are shown to be high, noisy and do not give contact location information, we have used the centroids of the output images from the tactile arrays to compare them.

Object Position

To verify that dynamics simulation of manipulation tasks is achievable, the interaction between the robot and the simulated environment must be evaluated. Towards this end, we have used the manipulated object position as a metric.

From the simulator, the object position is easily obtained given that the physics engine provides its center of mass each time step. However, in order to get the object position from the real execution, an object tracker is needed. The objects used in the experiments have been modelled to use the model-based object tracker from VISP [48]. This method is able to perform on-line object 6D tracking with enough accuracy to validate the simulator results. The object position is placed in one of its corners as shown in Fig. 4.30c. As the reference systems for the real and simulated object are not the same, the variation relative to the object starting position is used as the comparison metric.

4.6.3 Results

The analysis was performed from three different points of view: the execution results, the concrete quantitative results for each task and the global quantitative results that summarize how close the simulation was to reality.

4.6.3.1 Execution Results

The experiments were successfully carried out for the different manipulation tasks and pictures of the simulator and real executions can be seen in the Fig. 4.32. The task executions for both platforms were recorded and they can be seen in the video available at http://youtu.be/yCW3PDmIeOQ. Although the video is not showing perfect synchronization with real execution, it shows that the results of the tasks are very similar.

The executions in the real world lasted between 37 and 93 s. As the time step specified in the simulation needs to be small to get accurate results for the tactile sensors (0.0001 s), the time taken to execute the simulation experiments was significantly larger. They lasted between 10 and 30 min.

4.6.3.2 Quantitative Results

The results obtained for the real and simulated environment were compared using the metrics explained in Sect. 4.6.2.2. As the duration for both environments is very different, the times were recalculated as a percentage of the total duration time for each experiment.

Grasp

The five grasp experiments were executed successfully. As an example, detailed graphs for each metric are shown in Fig. 4.33 for the first grasp position. The arm and hand joints show very similar behaviour as the controller for the real and simulated environment is the same. In the tactile sensor graphs, it can be seen that the sensors of the robot start to get feedback before the simulated ones. However, when both are obtaining readings, the difference in texels is always less than 4 which allows the controller the use of simulated and real sensor feedback in a similar way. The position of the object recorded by the tracker has significantly more noise than the acquired with the simulator. However, the results are very similar in all axes, especially in z which was the direction of the object's movement. The rotation of the object in this experiments did not play a significant role which can be seen in the small variation of the plots.

Slide

The graphs showing the results obtained while sliding the pizza box can be seen in Fig. 4.34. The arm and hand joints gave similar results for both robots, as in the previous experiments, therefore their graphs were omitted. The tactile sensors in the case of the real robot, only show readings for the first sensor given that the hand was

(a)

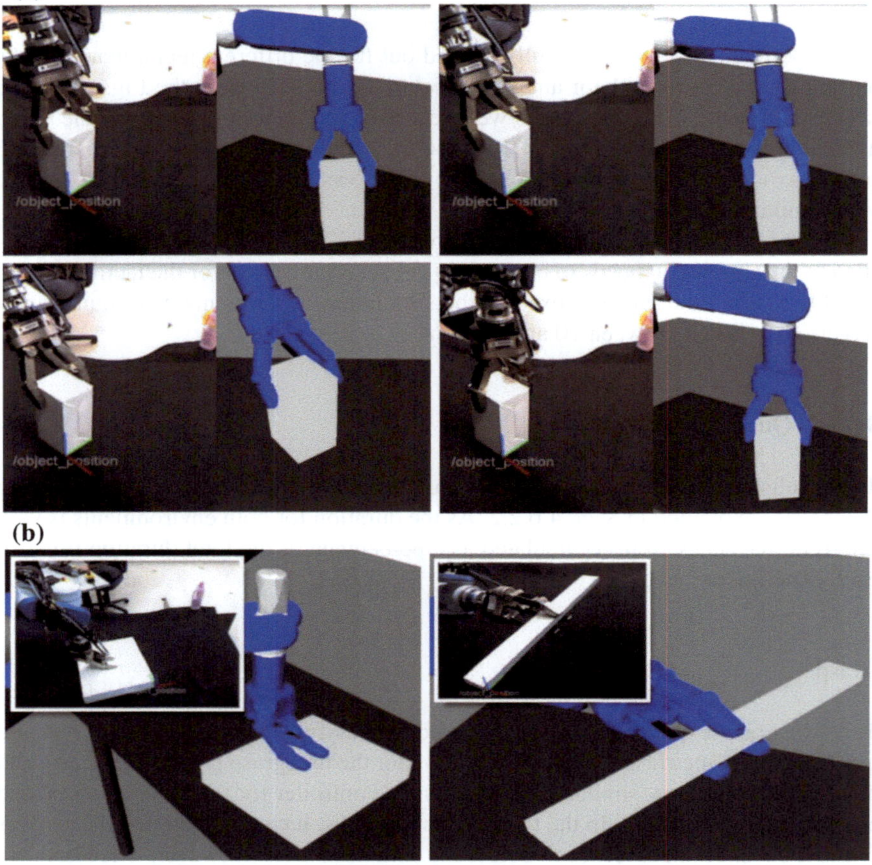

Fig. 4.32 Pictures of the real and simulated robot executing the manipulation tasks. **a** Grasping. **b** Sliding. **c** In-hand manipulation

slightly tilted towards that finger, then the pressure on the other tactile sensor was not enough to produce feedback.

However, in the simulated case both sensors gave similar readings. In addition, the real object is not rigid so it deformed when the fingers were in contact with the box surface, a behaviour that is not modelled with rigid objects in the simulator.

The object position shows a slight variation. The difference is explained by a lag which results from the simulated and real movements not occurring at the same time. This lag can be explained with the results from the tactile sensors, where it can be seen that the simulated sensor detected the contact before the real sensor which produced the sliding movement to start earlier in the simulation.

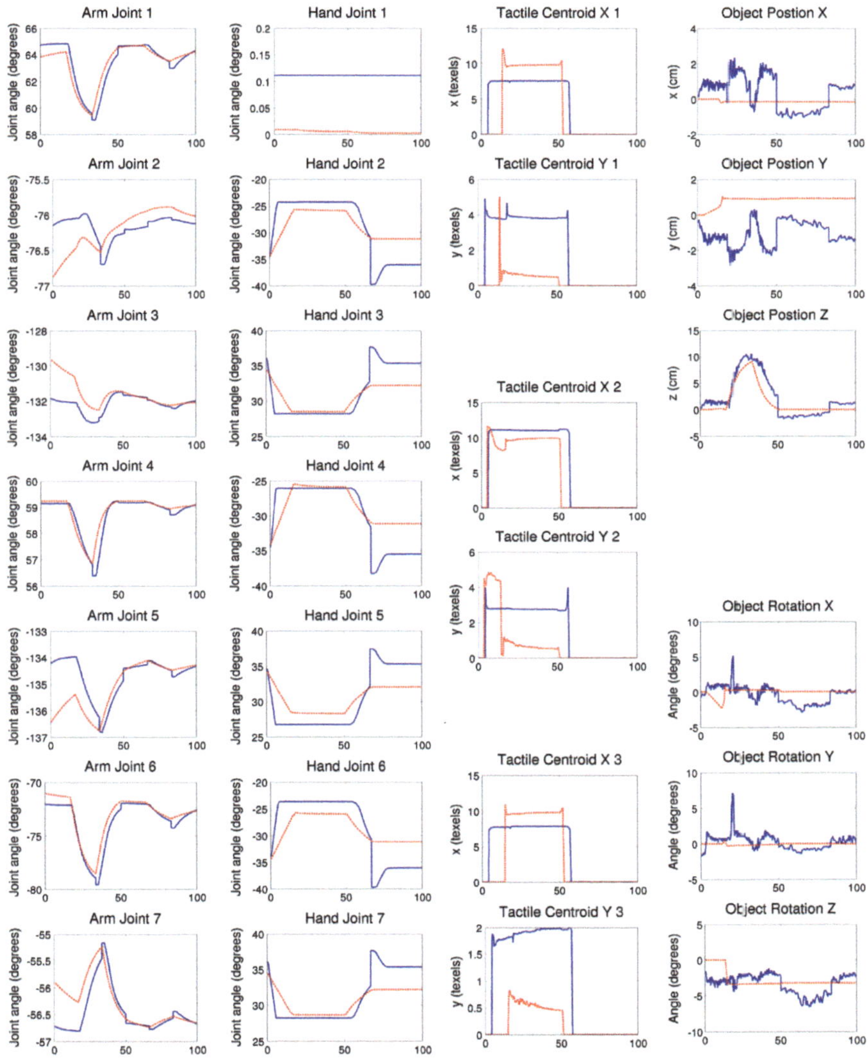

Fig. 4.33 Results for the first grasping experiment over time in percentage over the experiment duration. *Blue lines* indicate the readings from the robot and *red* ones from the simulator

With respect to the object rotation, although the box slides, the movement of the object is different. In the simulation, it rotates around 20° around its z axis, while there is no rotation in the real execution.

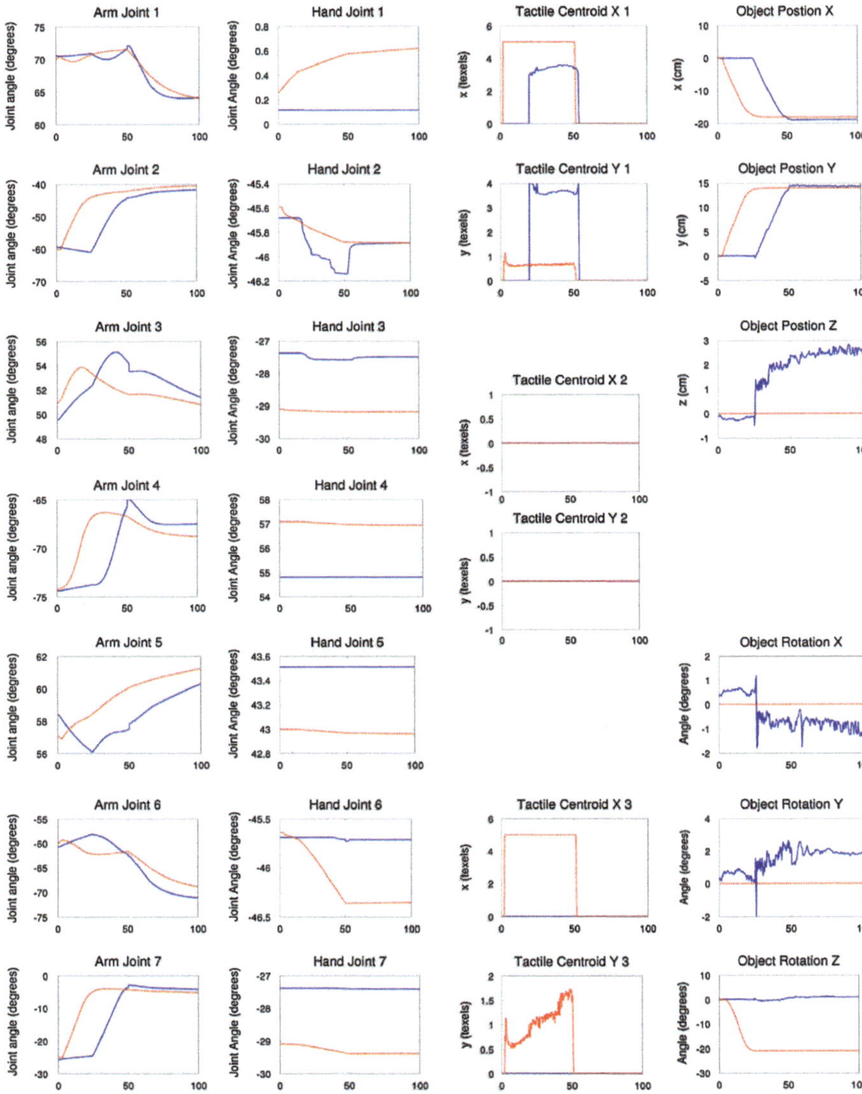

Fig. 4.34 Results for the sliding experiment over time in percentage over the experiment duration. *Blue lines* indicate the readings from the robot and *red* ones from the simulator

In-hand Manipulation

The results obtained by the in-hand manipulation of the wooden stick are shown in Fig. 4.35. The arm joints remained static during the experiment and hand joints showed similar behaviour as in the previous experiments, therefore both were omitted. The tactile readings show different results for the real and simulated case. The thumb finger tactile hardly shows any reading for both environments, while the other two sensors show higher readings in the simulated sensor than in the real one. This can be explained by the difference in the sensitivity of the sensors. Nevertheless, for this experiment the tactile readings were not used by the controller and the behavior was not influenced.

The object position and, specially the rotation in x, were the key measurements used to determine how close the simulation was to reality. The object position shows very different values. In the case of the simulation, it is near zero but for the real object it moves several centimeters. This can be explained by the difference in the object's reference system. While in the simulator, the center of the object remains more or less stable, in the real object the corner is the one that oscillates up and down with the hand movement.

However, the simulated object rotation in x, which was the most relevant in this experiment, shows a very similar behaviour to that of the real object. In the case of the simulation, the object was undergoing a rotation in y which was not the same as in reality, however the overall movement of the object remained very similar to the real one.

The errors and standard deviations were calculated for each of the metrics over all the experiments. The results can be seen in Fig. 4.36a. The arm joints presented the greatest error in the slide experiment. This can be due to the lag between the real and simulated executions that can be seen in Fig. 4.34.

The errors for the hand joints in Fig. 4.36b are in general higher than for the arm joints but inside an acceptable range of 0.1 radians. The hand-closing controller relies on tactile sensing, thus the detection of contacts is critical for the results to be equal. The different timing that the contacts have shown (see Figs. 4.33 and 4.34) may be the main cause for the hand joint error. An special case showed up in the Grasp experiment #2 where the distal joint of finger 1 did not detect contact with the object and continued moving while it was detected in the simulator execution.

The error in the object position (see dark blue column in Fig. 4.36c) is in general below 2.5 cm. Regarding the rotation error, it is below 0.05 radians which is very low. The special issues that rise up error, for the slide and in-hand manipulation task, are explained in Sects. 4.6.3.2 and 4.6.3.2.

Finally the differences in the tactile sensor readings are depicted in Fig. 4.36d. The mean error shows that although the tasks were completed successfully there were significant differences in the tactile sensor readings.

4.6.4 Discussion

In this section, we have presented a complete dynamics simulation of a humanoid torso robot. Moreover we have evaluated the extent to which the simulation resembles the real behavior of manipulation tasks by using the same controller on the real and the simulated platforms and analyzing the differences. The results have shown that

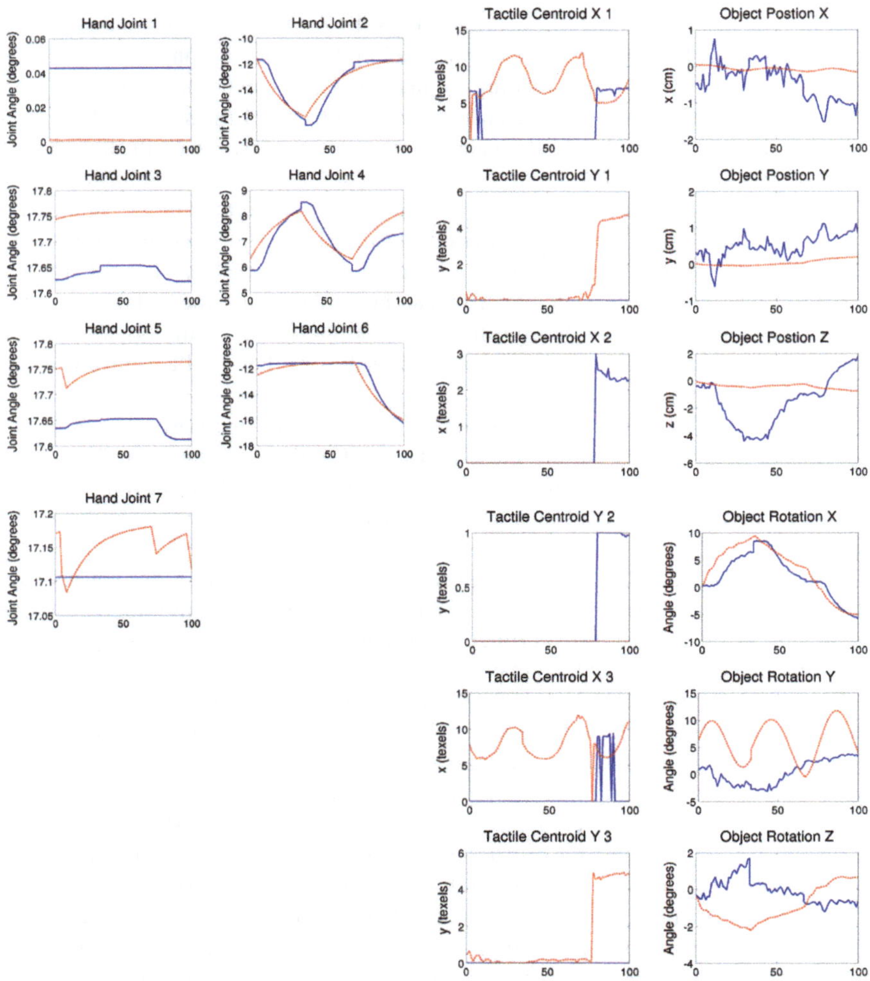

Fig. 4.35 Results for the in-hand manipulation experiment over time in percentage over the experiment duration. *Blue lines* indicate the readings from the robot and *red* ones from the simulator

it is possible to simulate manipulation tasks with the current state of the art of simulation tools. Although the precision is not perfect, the framework is able to perform manipulation tasks using the same controller as that used in the real world with very similar results. Therefore, an important result is that with this framework, it is possible to use simulated tactile data in manipulation task controllers that use tactile feedback.

The main drawback is the duration that the simulator requires to perform the same task as the real robot does, limiting its possible use as a prediction engine. However,

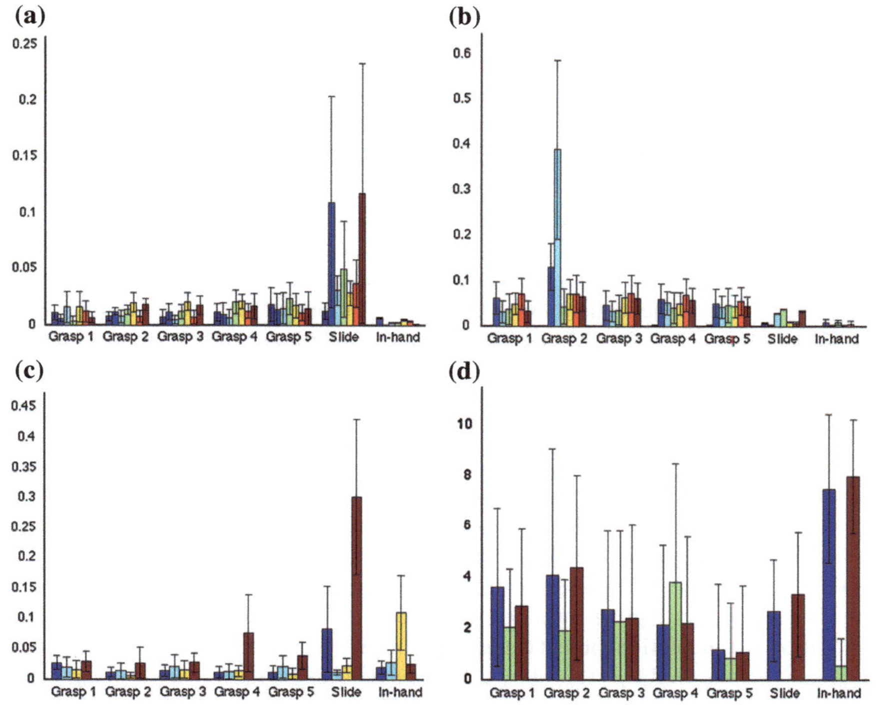

Fig. 4.36 Aggregated error for all the experiments performed. Each group of columns represents one experiment. **a** Arm joint errors (rad). Each column of a group represents the mean and stdev for each joint of the arm during the experiment execution. **b** Hand joint errors (rad). Each column of a group represents the mean and stdev for each joint of the hand during the experiment execution. **c** Object position errors for all the experiments. Each first column of a group represents the translation error in meters. The other three columns are rotation error in x, y, and z (rad). **d** Tactile sensor centroid position errors (texels). Each column of a group represents the position error of the detected centroid in texels for each finger

recent advances in parallelizing processes to improve the speed of simulations can be a feasible solution to this issue. The parallel quickstep package available in ROS provides an implementation in CUDA, OpenCL and OpenMP, which accelerates the process of calculating each time step. Another challenging problem was the parameter setting for the simulator, for which learning algorithms to find the right parameters could be used.

4.7 Conclusion

In this chapter, we presented several applications of the simulator towards achieving autonomous robot grasping. We used several tools described in the previous chapter which are included in the OpenGRASP framework. It is demonstrated how grasp simulation is a key tool for constructing a world model, understanding the robot environment and predicting robot actions.

Using the simulator to replicate the kinematics of the real platform, we show how it can be used to select an appropriate grasp and plan its trajectory in order to successfully grasp objects that are unknown or familiar to the robot. We presented the Model-Object Overlap Metric (MOOM) to incorporate information obtained from the depth-sensor into the grasp planning process and shown how it can be incorporated at different stages and in different fashions to conventional grasping pipelines to increase efficiency and robustness to errors from recognition.

For unknown objects, we reduced the search space for the optimal symmetry parameters through a good initialization, and we demonstrated the applicability of the predicted object meshes in a service robotic scenario by supporting execution in the real-world with grasp and motion planning in simulation.

Also, we presented a grasping pipeline that allows grasping according to an object's category and a given task. Several state-of-the-art modules performing scene exploration, segmentation, object categorization and task-based grasp selection were integrated. We showed how this allows the robot to transfer task-specific grasp experience between objects of the same category. The effectiveness of the pipeline was demonstrated on two humanoid robots.

Additionally, we demonstrated how to achieve a complete dynamic simulation of a humanoid robot using the developed toolkit. We have evaluated the extent to which the simulation resembles the real behavior of manipulation tasks by using the same controller on the real and the simulated platforms and analyzing the differences.

Although robots are still far away from being capable of human-level manipulation skills interacting in uncertain real-world environments, using the simulator as a prediction engine as demonstrated in this chapter, can help to achieve autonomous robot grasping.

References

1. Sahbani, A., El-Khoury, S., Bidaud, P.: An overview of 3d object grasp synthesis algorithms. Robot. Auton. Syst. **60**(3), 326–336 (2012), http://www.sciencedirect.com/science/article/pii/S0921889011001485, Autonomous Grasping
2. Bohg, J.: Multi-Modal Scene Understanding for Robotic Grasping. Ph.D. thesis, KTH Royal Institute of Technology (2011)
3. Bohg, J., Johnson-Roberson, M., Leon, B., Felip, J., Gratal, X., Bergstrom, N., Kragic, D., Morales, A.: Mind the gap - robotic grasping under incomplete observation. IEEE International Conference on Robotics and Automation (ICRA). pp. 686–693. Shanghai, China, (May 2011) http://ieeexplore.ieee.org/xpls/abs_all.jsp?arnumber=5980354

4. Bohg, J., Welke, K., Leon, B., Do, M., Song, D., Wohlkinger, W., Madry, M., Aldoma, A., Przybylski, M., Asfour, T., Marti, H., Kragic, D., Morales, A., Vincze, M.: Task-based Grasp Adaptation on a Humanoid Robot. In: 10th International IFAC Symposium on Robot Control (SYROCO 2012). Dubrovnik, Croatia (2012)
5. León, B., Felip, J., Marti, H., Morales, A.: Simulation of robot dynamics for grasping and manipulation tasks. 12th IEEE-RAS International Conference on Humanoid Robots (Humanoids) (2012)
6. Diankov, R.: Automated construction of robotic manipulation programs. Ph.D. thesis, Carnegie Mellon University, Robotics Institute, US (2010)
7. Miller, A., Allen, P.: Graspit! a versatile simulator for robotic grasping. Robot. Autom. Mag., IEEE **11**(4), 110–122 (2004)
8. Srinivasa, S., Ferguson, D., Vandeweghe, J.M., Diankov, R., Berenson, D., Helfrich, C., Strasdat, K.: The robotic busboy: Steps towards developing a mobile robotic home assistant. International Conference on Intelligent Autonomous Systems 2008
9. Srinivasa, S., Ferguson, D., Helfrich, C., Berenson, D., Collet, A., Diankov, R., Gallagher, G., Hollinger, G., Kuffner, J., VandeWeghe, M.: Herb: A home exploring robotic butler 2009, doi:10.1007/s10514-009-9160-9
10. Dogar, M., Srinivasa, S.: Push-grasping with dexterous hands: Mechanics and a method. In: Proceedings of 2010 IEEE/RSJ International Conference on Intelligent Robots and Systems (IROS 2010) (2010)
11. Klank, U., Pangercic, D., Rusu, R., Beetz, M.: Real-time cad model matching for mobile manipulation and grasping. In: Humanoid Robots, 2009. Humanoids 2009. 9th IEEE-RAS International Conference on. pp. 290–296 (2009)
12. Rusu, R., Holzbach, A., Diankov, R., Bradski, G., Beetz, M.: Perception for mobile manipulation and grasping using active stereo. In: Humanoid Robots, 2009. Humanoids 2009. 9th IEEE-RAS International Conference on. pp. 632–638 (2009)
13. Hsiao, K., Kaelbling, L., Lozano-Perez, T.: Robust grasping under object pose uncertainty. Autono. Robot. **31**, 253–268 (2011), http://dx.doi.org/10.1007/s10514-011-9243-2
14. Chiu, H.P., Liu, H., Kaelbling, L., Lozano-Perez, T.: Class-specific grasping of 3d objects from a single 2d image. In: Intelligent Robots and Systems (IROS), 2010 IEEE/RSJ International Conference on. pp. 579–585 (2010)
15. Fungtammasan, P., Watanabe, T.: Grasp input optimization taking contact position uncertainty into consideration. In: Robotics and Biomimetics (ROBIO), 2011 IEEE International Conference on. pp. 1326–1331 (2011)
16. Aldoma, A., Blodow, N., Gossow, D., Gedikli, S., Rusu, R.B., Vincze, M., Bradski, G.: CAD-model recognition and 6DOF pose estimation using 3D cues. In: Workshop: 3rd IEEE Workshop on 3D Representation and Recognition, ICCV Yet to appear, see http://3d-net.org/privatepapers/ICCVWorkshop.pdf (2011)
17. Wohlkinger, W., Vincze, M.: Shape distributions on voxel surfaces for 3D object classification from depth images. IEEE International Conference on Signal and Image Processing Applications (2011)
18. Goldfeder, C., Ciocarlie, M., Dang, H., Allen, P.K.: The columbia grasp database. IEEE International Conference on Robotics and Automation (2009)
19. Goldfeder, C., Ciocarlie, M., Peretzman, J., Dang, H., Allen, P.K.: Data-driven grasping with partial sensor data. IEEE/RSJ International Conference on Intelligent Robots and Systems. pp. 1278–1284 (2009)
20. Azad, P., Asfour, T., Dillmann, R.: Stereo-based 6D object localization for grasping with humanoid robot systems. IEEE/RSJ International Conference on Intelligent Robots and Systems (IROS). pp. 919–924 (2007)
21. Brook, P., Ciocarlie, M., Hsiao, K.: Collaborative grasp planning with multiple object representations. IEEE International Conference on Robotics and Automation, ICRA (2011)
22. Xue, Z., Kasper, A., Zoellner, M.J., Dillmann, R.: An automatic grasp planning system for service robots. In: Proceedings of the 14th International Conference on, Advanced Robotics (2009)

23. Aldoma, A., Vincze, M.: Pose alignment for 3D models and single view stereo point clouds based on stable planes. International Conference on 3D Imaging, Modeling, Processing, Visualization and Transmission (2011)
24. Felip, J., Morales, A.: Robust sensor-based grasp primitive for a three-finger robot hand. IEEE/RSJ International Conference on Intelligent Robots and Systems, 2009. IROS 2009. pp. 1811–1816 (2009)
25. Breckon, T.P., Fisher, R.B.: Amodal volume completion: 3d visual completion. Comput. Vis. Image Underst. **99**(3), 499–526 (2005)
26. Thrun, S., Wegbreit, B.: Shape from symmetry. Computer Vision, IEEE International Conference on **2**, 1824–1831 (2005)
27. Kazhdan, M., Bolitho, M., Hoppe, H.: Poisson surface reconstruction. In: SGP '06: Proceedings of the fourth Eurographics symposium on Geometry processing. pp. 61–70. Eurographics Association, Aire-la-Ville, Switzerland, Switzerland (2006)
28. Ratcliff, J.: Convex decomposition library (2009), http://codesuppository.blogspot.com/2009/11/convex-decomposition-library-now.html
29. Richtsfeld, M., Vincze, M.: Grasping of unknown objects from a Table top. In: ECCV Workshop on 'Vision in Action: Efficient strategies for cognitive agents in complex environments'. Marseille, France (2008)
30. Papazov, C., Burschka, D.: Stochastic optimization for rigid point set registration. In: ISVC '09: Proceedings of the 5th International Symposium on Advances in Visual Computing. pp. 1043–1054. Springer, Berlin (2009)
31. Roy, M., Foufou, S., Truchetet, F.: Mesh comparison using attribute deviation metric. Int. J. Image Graph. **4** (2004)
32. Song, D., Huebner, K., Kyrki, V., Kragic, D.: Learning task constraints for robot grasping using graphical models. IEEE/RSJ International Conference on Intelligent Robots and Systems 2010
33. Song, D., Ek, C.H., Huebner, K., Kragic, D.: Embodiment-specific representation of robot grasping using graphical models and latent-space discretization. IEEE/RSJ International Confrence on Intelligent Robots and Systems. pp. 980–986 (2011)
34. Marton, Z.C., Pangercic, D., Blodow, N., Beetz, M.: Combined 2d–3d categorization and classification for multimodal perception systems. Int. J. Robot. Res. (2011), http://ijr.sagepub.com/content/early/2011/07/23/0278364911415897.abstract
35. Fritzke, B.: A growing neural gas network learns topologies. In: Advances in Neural Information Processing Systems 7. pp. 625–632. MIT Press (1995)
36. Lowe, G.D.: Distinctive image features from scale-invariant keypoints. Int. J. Comput. Vision **60**(1), 91–110 (2004)
37. van de Sande, K.E.A., Gevers, T., Snoek, C.G.M.: Evaluating color descriptors for object and scene recognition. Patt. Anal. Mach. Intell. **32**(9), 1582–1596 (2010)
38. Przybylski, M., Asfour, T., Dillmann, R.: Planning grasps for robotic hands using a novel object representation based on the medial axis transform. IEEE/RSJ International Conference on Intelligent Robots and Systems 2011
39. Blum, H.: Models for the perception of speech and visual form, chap. A transformation for extracting new descriptors of shape, pp. 362–380. MIT Press, Cambridge, Massachusetts (1967)
40. Miklos, B., Giesen, J., Pauly, M.: Discrete scale axis representations for 3d geometry. In: ACM SIGGRAPH. ACM, New York (2010)
41. Ferrari, C., Canny, J.: Planning optimal grasps. In: Proceedings 1992 IEEE International Conference on Robotics and Automation pp. 2290–2295 (1992)
42. Madry, M., Song, D., Kragic, D.: From object categories to grasp transfer using probabilistic reasoning. IEEE International Conference on Robotics and Automation (2012) (to appear)
43. Pastor, P., Righetti, L., Kalakrishnan, M., Schaal, S.: Online movement adaptation based on previous sensor experiences. IEEE/RSJ International Conference on Intelligent Robots and Systems. pp. 365–371. San Francisco, USA (2011)

44. Laue, T., Hebbel, M.: Automatic parameter optimization for a dynamic robot simulation. In: Iocchi, L., Matsubara, H., Weitzenfeld, A., Zhou, C. (eds.) RoboCup 2008: Robot Soccer World Cup XII, Lecture Notes in Computer Science, vol. 5399, pp. 121–132. Springer, Berlin (2009)
45. Weitnauer, E., Haschke, R., Ritter, H.: Evaluating a physics engine as an ingredient for physical reasoning. In: Proceedings of the Second international conference on Simulation, modeling, and programming for, autonomous robots. pp. 144–155 (2010)
46. Zhang, L., Betz, J., Trinkle, J.: Comparison of simulated and experimental grasping actions in the plane. First Joint International Conference on Multibody System Dynamics 2010
47. León, B., Ulbrich, S., Diankov, R., Puche, G., Przybylski, M., Morales, A., Asfour, T., Moisio, S., Bohg, J., Kuffner, J., Dillmann, R.: OpenGRASP: A toolkit for robot grasping simulation. In: Simulation, Modeling, and Programming for Autonomous Robots, Lecture Notes in Computer Science, vol. 6472, pp. 109–120. Springer, Berlin (2010)
48. Marchand, E., Spindler, F., Chaumette, F.: Visp for visual servoing: a generic software platform with a wide class of robot control skills. IEEE Robot. Autom. Mag. **12**(4), 40–52 (2005) (Special Issue on "Software Packages for Vision-Based Control of Motion")

Part II
Human Grasping Simulation

Chapter 5
The Model of the Human Hand

5.1 Introduction

Most of human mechanical interactions with the surrounding world are performed by the hands. They allow us to perform very different tasks; from exerting high forces (e.g. using a hammer) to executing very precise movements (e.g. cutting with a surgical tool). This versatility is possible because of a very complex constitution: a great number of bones connected through different joints, a complicated musculature and a dense nervous system. This complexity is already evident from the kinematics point of view, with more than 20° of freedom (DOF) controlled by muscles, tendons and ligaments.

Mathematical representations are used in order to perform qualitative or quantitative analyses on this complex reality. These representations are known as biomechanical models of the hand. In biomechanics, their use allows studying problems that cannot be analysed directly on humans or that have an experimental cost that is too high; e.g. the study of new alternatives for restoring hand pathologies. Biomechanical models are a description of the hand as a mechanical device: the different elements of the hand are defined in terms of rigid bodies, joints and actuators, and the mechanical laws are applied. As they are simplified mathematical models of the physical entity, their use and validity depend on the simplifications considered.

While one of the main features of the human hand is its grasping capability, the current models have a very limited ability for its simulation. All the effort in biomechanics has been focused on appropriately modelling the different hand components (kinematics, muscles, tendons, etc.). Little effort has been spent on the formulation of the grasping problem when using a biomechanical model. In this sense, many limitations persist. Current models do not allow the estimation of the contact information required to use biomechanical models for simulating the grasping of objects. Forces and zones of contact still need to be measured experimentally and input to the model.

Robot hand grasps have been extensively studied for years. Although until 2000 little attention was paid to human hand grasping, this too has become a hot topic in

B. León et al., *From Robot to Human Grasping Simulation*,
Cognitive Systems Monographs 19, DOI: 10.1007/978-3-319-01833-1_5,

robotics. In this sense, although the human hand is obviously more complex than robot hands, the methods used in robotics might be adopted to study the human grasp by considering the hand as the human end-effector.

Furthermore, much research has been carried out on animation techniques over the past years, mainly for use in developing computer games. Lately, these advances have been cleverly used by some ergonomics researchers to develop improved graphical and kinematics hand models for evaluating the use of products [1–4], with good results.

A promising research area lies ahead with scientist, aiming to obtain a more comprehensive model of the hand, integrating knowledge and developments from the fields of biomechanics, ergonomics, robotics, and computer animation.

5.2 Literature Review

5.2.1 Biomechanical Models of the Hand

Over the years, biomechanical models of the hand have been developed for different purposes. Some of them tried to study the functionality of different anatomical elements with the aim of gaining a deeper understanding of the causes and effects of many hand pathologies. These are usually very simplified (mostly two-dimensional) kinematic models (sometimes dynamic) that are used to perform qualitative analyses [5, 6]. Others were developed to help in medical planning and surgery for patients; they are usually dynamic models and are used to perform quantitative analyses, such as the study of the tendon excursions in the medical planning of tendon transfers [7] or to study the nerve stimulation required to restore the grasping ability in muscular dysfunction patients [8]. Yet others studied the hand while performing specific tasks with different aims, so as to have approximate values for the articular forces for testing prosthetic designs [9]. These too are quantitative analyses performed on dynamic models.

Recent models do not differ much from the ones developed before 2000 [10–26]. All models present a similar configuration. The kinematics are modelled without considering the restraining structure, just the resultant physiological articular movement. The concept of the instantaneous centre of rotation has been used to define an axis of rotation in joints with a single predominant DOF. Much effort has also been spent on finding the rotation axes of joints with two DOF [27], through the consideration of a virtual link connecting the axes [28]. Thus, all works use fixed axes of rotation; depending on the joint, one or two axes of rotation are considered. This approximation has been found to be good enough for most of the cases, particularly if there is no interest in analysing the role of the articular soft tissue or the articular stresses [29].

All works in the literature consider the ideal case of a non-friction belt around a pulley to model the tendons crossing the joints. Therefore, the tensional force on a tendon is the same along its pathway if no split or connection to other tendons exists. The most-used approach to model tendon action on the joints considers the tendon freely running when crossing the joint between two points attached one to the proximal segment of the joint and the other to the distal segment. This approach is the basis of the first serious attempt to develop a 3D normative model of the hand [30], in which the position of the tendons with respect the bone segments were obtained from the measurement of 10 fresh cadaveric specimens.

Most of the works in the literature use Hill's model to account for the muscles' mathematical modelling. This simple model allows the consideration of the three main parameters, i.e. muscle activation level and variation of the maximum deliverable muscle force with muscle length and muscle contraction velocity.

Finally, the dynamic equilibrium equations lead to an indeterminate system of equations, with more unknowns (muscle forces) than available equations. Inequality constraints taking into account the maximal forces that may be delivered by each muscle and that tendons cannot support compressive forces have to be considered as well. The problem is usually solved by minimising some cost function. Different functions have been investigated, most of them without any physiological basis. The most often used criterion is the minimisation of the sum of the squared muscle stresses, which has been related to the maximisation of fatigue resistance [31].

5.2.2 Hand Models in Ergonomics

Ergonomics, according to the International Ergonomics Association, is "the scientific discipline concerned with the understanding of interactions among humans and other elements of a system, and the profession that applies theory, principles, data and methods to design in order to optimise human well-being and overall system performance". Hand models in ergonomics are used to simulate postures adopted while grasping objects with different purposes. One of the main goals of physical ergonomics is the study of the size and shape of objects according to the anthropometry of the different people that have to interact with them. Thus, the main feature of a model for ergonomics is that it has to allow representing different populations and percentiles. People having hands of different sizes and proportions will adopt different postures in grasping the same object for the same functions. For example, pressing a button on a phone with the thumb while holding it with the same hand can be easily achieved for a specific hand size while maintaining the grasp. However, other people with different a size of hand will need to change the grasping posture to press the button. This is a typical problem of reach that needs to be solved in ergonomic assessment.

In recent years, virtual humans have been incorporated into the design process for ergonomic assessment of different types of products, mainly in the aerospace and automotive industry but also in others like product design, task simulation, personnel training or simulation of other worker environments [32, 33]. Several commercial software programs such as Jack, RAMSIS, HumanCAD, Safework and

SantosHuman are available and other studies have been conducted on digital human models such as SAMMIE [34] or the Boeing Human Modeling System for the same purposes. A virtual human in these packages is defined as a kinematic chain composed of a number of rigid links connected by joints. These joints have the DOF and allowable motion limits corresponding to the anatomical joint of the human being. Direct and inverse kinematics is incorporated into the models so they can replicate human body movements and also evaluate forces acting on joints. Moreover, different populations and percentiles may be selected for the size of the model, usually from known anthropometric databases. With these capabilities the problems of reach and clearance typical in ergonomics may be solved easily. Other useful capabilities of these models are the simulation of the sense of sight with virtual cameras located in the eyes or the possibility to change any particular parameters of the model, like dimensions of limbs or motion limits of some joints, in order to simulate a particular person or disability. However, the majority of these models focus on the whole body and do not pay attention to the accuracy of the hand model. Most of them just incorporate a list of hand postures (grasping or others) to be chosen, i.e. direct kinematics, but do not allow for example inverse kinematics for the joints of the hand, even when it is incorporated for the other joints of the body. In recent developments some attempts to improve the hand model incorporated into some programs have been made [35, 33].

Another important aspect of hand models for ergonomics is associated with the study of musculoskeletal disorders. Early epidemiological studies [36] showed that the use of hand tools with an improper design for the worker or the task could lead to a high risk of developing cumulative hand trauma disorders (CHTD). The factors influencing the development of CHTD have been reported in different works [37–41] and different methods have been used in these studies: epidemiological studies, physiological measurements (electromyography activity, pressure in tissues, posture of hand and wrist, tactile sensitivity), biomechanical models of hand and wrist structures and psychophysical assessments. These studies report that CHTD are associated with repetitive tasks, high forces, extreme or awkward postures of hand and wrist, velocity and acceleration of wrist motions and exposure time, among others. Different theories of injury development have been proposed [42]. All of them assume that CHTD and other musculoskeletal disorders are of biomechanical nature. Therefore, biomechanical hand models able to predict movements, postures and internal forces of hand and wrist structures can be used to assess the risk of developing CHTD. Tendon excursions or maximum gripping strength have been used as indices in different works to assess gripping posture for health [20, 43].

None of the reported biomechanical models of the hand for ergonomics accounts for all requirements, although some attempts have been made. [43] have developed a scalable kinematic model of the hand with simple geometry (cones and cylinders). The model includes a posture prediction algorithm for fingers that reproduces in a high percentage the angles of the observed postures and is able to compute tendon excursions and wrist movements. The model is used to assess how much space is required for hands in an assembly task and to calculate the risk of CHTD from tendon forces and hand strength. Another group of researchers [1, 2, 4] have developed a scalable digital hand model with an accurate shape of the hand that includes a semi-automatic grasp planning function with robotics indeces of quality. The model incorporates a "comfort database" obtained from experimental measurements to assess comfort of postures and is used in the assessment of physical interaction with electronic appliances.

5.2.3 Grasping in Robotics

For many years the robotics community has been studying the autonomous handling of objects by robots, which is the main focus of the first part of this book. As presented in Chap. 2, a grasp is commonly defined as a set of contacts on the surface of the object. A contact model should be defined to determine the forces or torques that the robot manipulator must exert on the contact areas. The force applied by a finger at a contact point generates a *wrench* on the object with force and torque components. The contact model maps the wrench at some reference point of the object, usually the centre of mass. The most common contact models used in robotic grasping are the point contacts with and without friction and the soft-finger contacts (see Sect. 2.2.2).

After establishing the contact model, it can be used to study tasks involving multiple contacts. The set of contacts defining each grasp can be analysed in order to test its ability to resist disturbances and its dexterity properties. The grasps that can be maintained for every possible disturbing load are known as *closure grasps*. However, there is usually more than one grasp that fulfils this condition. Many grasp quality metrics and approaches have been proposed to evaluate the dexterity of the selected grasps and determine which one is the best to be executed. They are reviewed in the following chapter.

Two main problems can be distinguished in robotic grasping: analysis and synthesis. Grasp analysis (Sect. 2.3) consists of determined whether the grasp is stable using common closure properties, given an object and a set of contacts. Then, a quality measure can be evaluated in order to enable the robot to select the best grasp to execute. On the other hand, given an object, grasp synthesis algorithms should provide a suitable set of contacts on the object surface and determine an appropriate hand configuration (Sect. 2.4). Usually they take the geometry of the object as an input to select optimal force-closure contact locations or whole regions that yield force closure. These contacts are the starting point for grasp analysis and dexterous manipulation methods.

Despite many years of research and all the advances we have reviewed, the robotics community is still not able to build a manipulator with similar capabilities to the human hand. The robot hands constructed until now are only simplifications, given the complexities not only at the sensor and actuator level, but also at the control level.

5.3 Hand Model Proposed for the Study of Grasp

Based on the literature review, current hand biomechanical models allow estimating the muscular patterns required to perform a movement while counteracting a system of external forces. But their use for studying object grasping is limited. On the one hand, biomechanical models lack realism for assessing the use of handheld products from an ergonomics point of view. Hand models in ergonomics have reached a high level of realism but do not allow for mechanical analyses. On the other hand, biomechanical models are not self-contained, as they need contact information to be input to the model. Current models do not allow predicting grasping postures nor evaluating contact forces and zones, much less predicting the movements while grasp planning. Quality grasp measures in robotics allow comparing different robotic grasping postures and could be adapted to human grasping.

A detailed description of our approach used to model the different components of the hand is provided in the following sections: joints-kinematics, muscles, ligaments and passive tissues, skin, contact with objects and neuromuscular control. The features that we require in order to create a model are:

- The model has to simulate the complete hand in order to allow the study of any grasp.
- The model has to be scalable to allow the simulation of different population groups.
- The model has to simulate and show the grasping of an object in a realistic way.
- The model has to estimate the muscular patterns required to perform a movement while counteracting the system of external forces that define the object manipulation. Furthermore, the model has to estimate the articular forces at the hand joints.
- The model has to be dynamic in order to allow the study of any grasping task (slow or fast) during the object manipulation process.
- The model has to predict feasible grasping postures for a given object and provide the contact information required for evaluating the grasp.
- The model has to incorporate quality grasping measures for evaluating the grasp.

5.4 Anatomy of the Hand: Terminology

In this section, a brief description of the anatomy of the hand is presented in order to lay down the basic terminology for the following sections describing the biomechanical model.

The human hand has 27 bones divided into three groups: 8 *carpal* bones in the wrist, 5 *metacarpal* bones, and 14 *phalanges* in the fingers [44]. There are three phalanges for each digit and two for the thumb, which are labeled *proximal*, *middle* and *distal* phalanges (with the middle one missing in the case of the thumb), according to their positions and become progressively smaller. There are four jointsin each

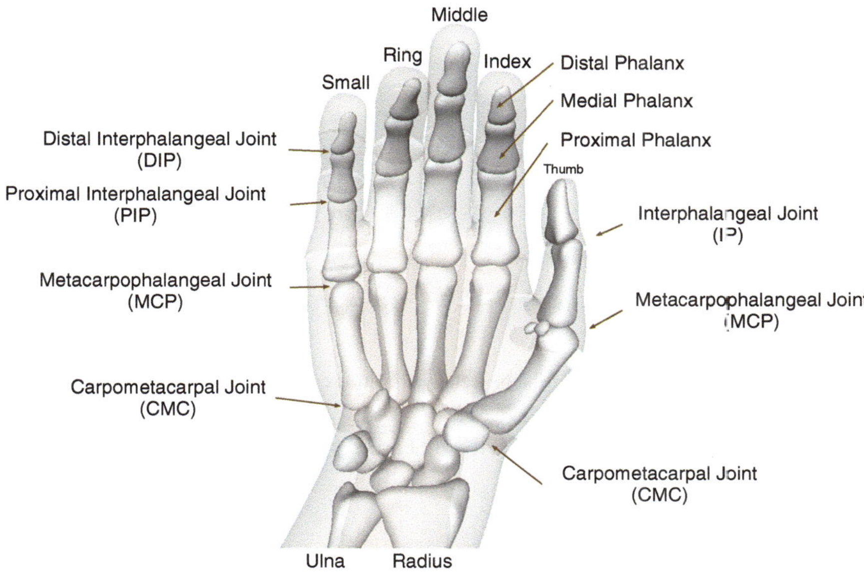

Fig. 5.1 Bones and joints of the *right* hand (*palmar view*)

finger, in sequence from the proximal to distal: *carpophalangeal* (CMC), *metacarpophalangeal* (MCP), *proximal interphalangeal* (PIP) and *distal interphalangeal* (DIP) joints (see Fig. 5.1).

The muscles producing movement of the fingers are divided into extrinsic and intrinsic based on the origin of the muscles. The *extrinsic* muscles originate primarily in the forearm, while the *intrinsic* muscles originate primarily in the hand. The extrinsic muscles are divided into *flexors* found primarily on the anterior forearm and *extensors* found primarily on the posterior forearm. Both set of muscles insert on the carpal bones, metacarpal or phalanges. The intrinsic muscles are divided into three groups: the thenar, the hypothenar and the midpalmar muscle groups. Details of all these muscles of the hand can be found in [44].

The standard terminology to identify the relative position of the hand elements, according with the three spacial directions, is shown in Fig. 5.2. Additionally, the nomenclature used to describe the different movements that the hand can perform are graphically described in Fig. 5.3.

5.5 Biomechanical Model

A previously-validated 3D, scalable, biomechanical model of the complete hand [18–21] has been adapted. The existing biomechanical model was developed in a scalable way, choosing two well-known anthropometric parameters of the hand: the

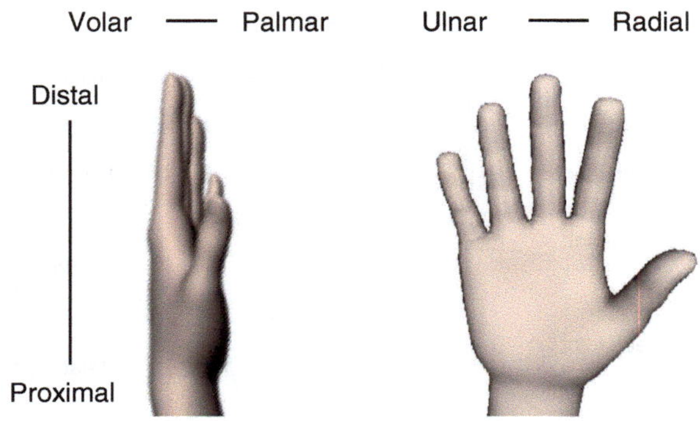

Fig. 5.2 Nomenclature used to describe the relative positions in the three spatial directions

hand length (HL) and hand breadth (HB) that are easy to measure and representative

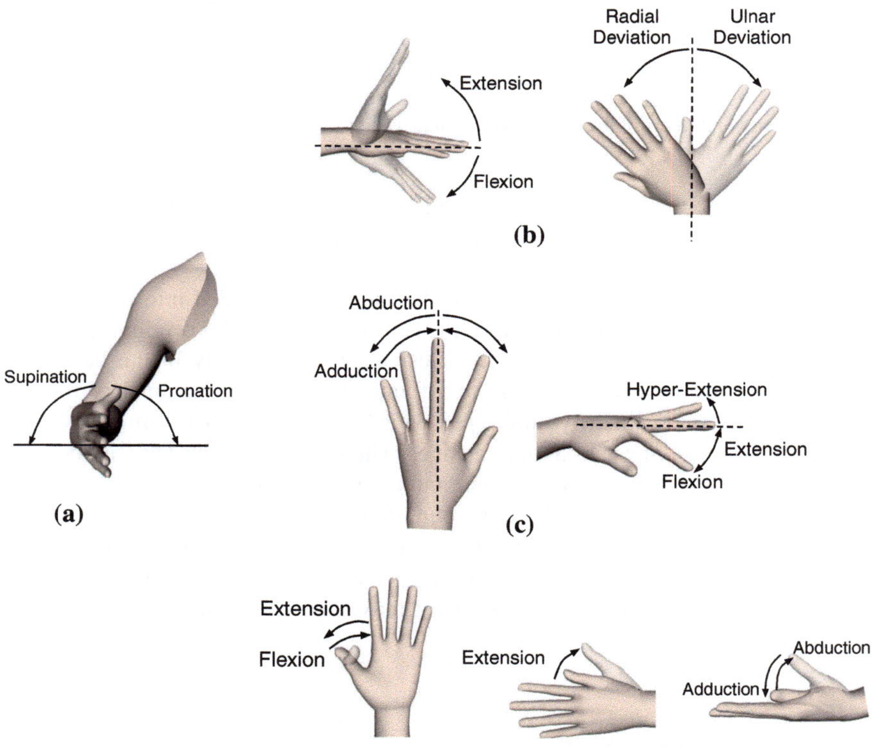

Fig. 5.3 Movements of the different elements of the hand. **a** forearm, **b** wrist, **c** fingers, **d** thumb

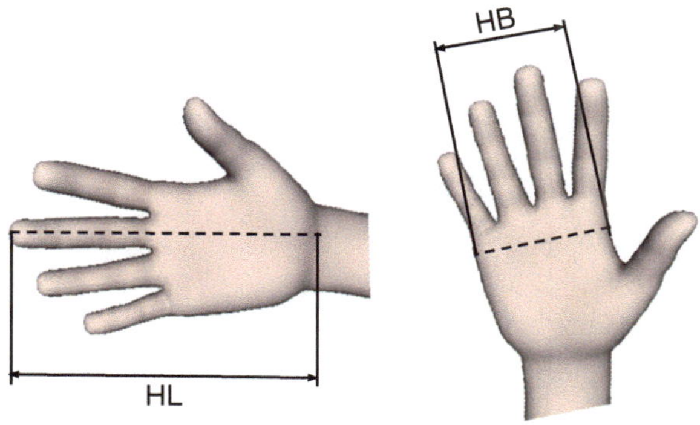

Fig. 5.4 Parameters used to scale the model: HL (*hand length*) and HB (*hand breadth*)

of the hand size. The parameters are shown in Fig. 5.4. It enable us to define the kinematics of the hand according with the hand parameters (Sect. 5.5.1) and to estimate the muscular patterns required to perform movements while counteracting external forces (Sects. 5.5.2 and 5.5.3). But in this work, we have adapted this model to use it for studying object grasping. We added a contact model (Sect. 5.5.4) to obtain the contact information needed as input to the evaluation of contact and muscle forces required for grasping an object. We have also implemented several grasp quality measures, which also use this contact information to compare different grasping postures. This is a major part of our work, therefore it is described in detail in the following chapters (see Chaps. 6 and 7). Additionally, we have also added a realistic geometry of the skin (Sect. 5.5.5) in order to more accurately provide the required contact information for evaluating grasp.

The model was implemented on the platform used for robotic grasp simulation described in the first part of this book. The model uses the robotics platform capabilities to perform the analysis of grasping (see Sect. 5.6).

5.5.1 Kinematics

The coordinate systems used to define the model were chosen following the ISB recommendations [45] for the shoulder, elbow, wrist and hand (see Fig. 5.5). The hand model considers 23° of freedom (DOF) selected to realistically simulate the hand movements. The hand has been considered as five skeletal open chains of rigid bodies connected to the carpus through different joints which characterise the kinematic behaviour of the chains.

Distal and proximal interphalangeal (DIP and PIP) joints of the fingers as well as the interphalangeal (IP) joint of the thumb are trochlear joints, capable only of flexion/extension movements [27]. These joints are modelled as one DOF joints by means of defining a rotation axis connecting the adjacent phalanges (hinge joint).

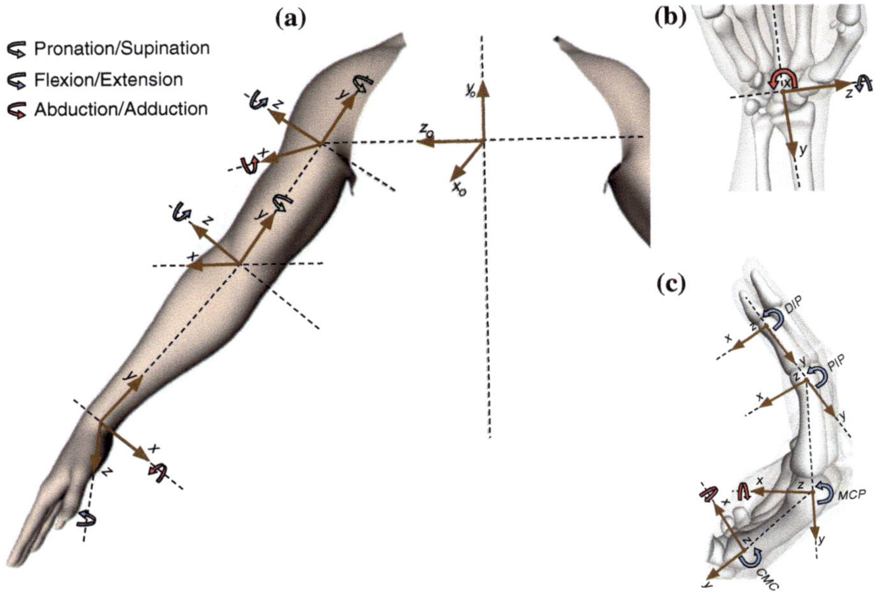

Fig. 5.5 Coordinate systems and definition of motions: (**a**) Global and arm coordinate systems, (**b**) wrist coordinate system and (**c**) finger coordinate systems

Thumb and finger metacarpophalangeal (MCP) joints are condylar joints, capable of flexion/extension and abduction/adduction movements [27]. The thumb carpometacarpal (CMC) joint is a saddle joint, capable also of flexion/extension and abduction/adduction movements [27]. All these joints are modelled as two DOF joints by defining two axes of rotation connecting the adjacent segments. In reality, the axes are neither intersecting nor orthogonal [27], so that a virtual link can be used to connect both axes [28]. However, these joints have been modelled as universal joints with orthogonal and intersecting axes for simplicity. This is an important aspect to improve in future work.

Finally, the hand model allows the arching of the palm by modelling the CMC joints of the little and ring fingers. These joints are arthrodial joints, with a very limited range of movement [46]. They have been modelled as one DOF joints by means of defining a flexion/extension axis of rotation connecting the carpus to each metacarpal. Due to the important role that the shape of the palm plays in grasping, this model is considered more suitable for grasping simulation than others in the literature.

The data for the location and orientation of the rotation axes comes from [30, 47, 48]. Axes data and link lengths are fully scaled with respect to the hand length and hand breadth [49]. Upper and lower limits for the joints have been obtained from [50, 51] and are shown in Table 5.1.

Table 5.1 Angles defining the upper and lower limits for each joint (°)

Fingers	CMC Flex.		CMC Abd.		MCP Flex.		MCP Abd.		PIP Flex.		DIP Flex.	
	−	+	−	+	−	+	−	+	−	+	−	+
Thumb	−25	35	−30	60	−10	80	−30	60	−15	80	–	–
Index	–	–	–	–	0	90	−15	42	0	100	−10	90
Middle	–	–	–	–	0	90	−8	35	0	100	−10	90
Ring	0	15	–	–	0	90	−20	14	0	100	−20	90
Small	0	30	–	–	0	90	−40	19	0	100	−30	90

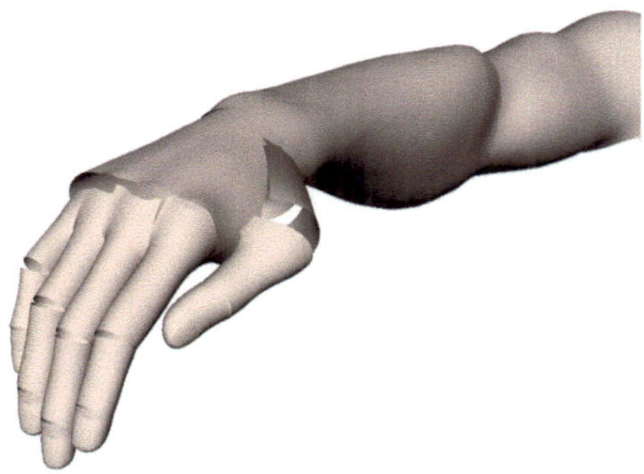

Fig. 5.6 Relaxed human hand posture

In order to study the forward and inverse kinematics of the hand, the Denavit–Hartenberg method from the robotics field [52] was adapted to define the position of any segment point.

5.5.1.1 Relaxed Human Posture

The position in which all muscles of the hand do not work is known as the *hand relaxed posture*. This posture is useful to measure the comfort of a given grasp (see Sect. 6.3.4). [53] experimentally determined different relaxed hand postures varying the pronation and flexion degrees in the shoulder joint. We used the results of the hand posture when the shoulder is at 0° in pronation and flexion which is the most similar to the grasp postures used later in this work. The angles in degrees used to define this posture (see Fig. 5.6) are shown in Table 5.2.

5.5.2 *Musculo-tendon Action*

Muscles and tendons control the movement of the skeletal chains. Muscles have been considered using a simple Hill three-component model (Hill 1938) that takes

Table 5.2 Angles for the relaxed hand posture (°)

Fingers	CMC Flex.	CMC Abd.	MCP Flex.	MCP Abd.	PIP Flex.	DIP Flex.
Thumb	0.0	0.0	46.1	0.0	8.5	–
Index	–	–	28.4	0.0	25.5	13.1
Middle	–	–	32.8	0.0	30.1	12.7
Ring	2.0	–	24.7	0.0	34.5	11.7
Small	5.0	–	16.6	0.0	32.1	15.6

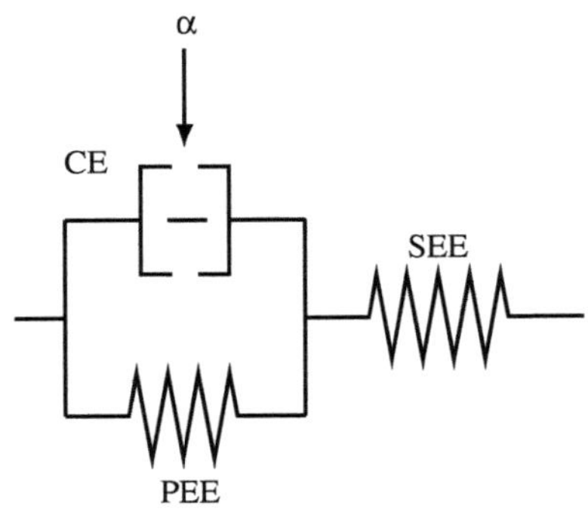

Fig. 5.7 Hill's three-component model for the muscles

into account the muscle activation level (α) and the force-length and force-velocity relationships, as well as the different index of architecture of muscles. The model considers a contractile element (CE), which is the basic component that generates force, a parallel elastic element (PEE), which is responsible for the passive force generated by the muscle when it is stretched, and a series elastic element (SEE), the muscle tendon unit, which has been considered to be inextensible (Fig. 5.7).

The force a muscle can exert depends on the actual muscle length and contraction velocity. It is widely accepted [54] that the maximum force a muscle can exert in optimal conditions is proportional to its physiological cross-sectional area (PCSA):

$$F_{max} = PCSA \cdot S_{max} \tag{5.1}$$

where S_{max} is the maximum stress the muscle can bear, which has been considered the same for each muscle [54].

The strain of tendons is insignificant for the magnitude of forces developed by the muscles [55]. Under this consideration, the SEE has been considered to be inextensible, so that the force the muscle exerts (F) can be written as:

$$F = F_{max}(F_{CE} + F_{PEE}) \tag{5.2}$$

where F_{CE} and F_{PEE} are the normalised forces delivered by the CE and PEE, respectively.

The force exerted by the muscle can be decomposed into an active and a passive force corresponding to the forces delivered by the CE and PEE, respectively. The force delivered by the CE is related to the muscle architecture and is a function of the muscle length l_{CE}, the contraction velocity v_{CE}, and the muscle activation level α (from 0 to 1), which is controlled by the central nervous system [56]:

$$F_{CE} = \alpha \cdot F_l(l_{CE}) \cdot F_v(v_{CE}) \tag{5.3}$$

where F_l and F_v are the non-dimensional force-length and force-velocity relationships.

A characteristic bell-shaped curve exists between force and length of the muscle. To model this dependence, the expression proposed by [56] has been used:

$$F_l(\epsilon, i_a) = e^{-\left[\frac{(\epsilon+1)^{0.96343\cdot(1-\frac{1}{i_a})}-1.0}{0.35327\cdot(1-i_a)}\right]^2} \text{ for } i_a < 1 \tag{5.4}$$

$$F_l(\epsilon, i_a) = e^{-[2.727277\cdot ln(\epsilon+1)]^2} \text{ for } i_a = 1 \tag{5.5}$$

where i_a is the muscle architecture index, defined as the ratio between the muscle fibre length and the muscle belly length, and ϵ is the muscle strain due to its lengthening from l_o, the muscle length for the optimal conditions.

The force a muscle can exert decreases when the contraction velocity of the muscle fibres increases. To model this dependence the expression proposed by [57] has been used

$$F_v(\dot{\eta}) = \frac{0.1433}{0.1074 + e^{-1.409\cdot sinh(3.2\cdot\dot{\eta}+1.6)}} \tag{5.6}$$

where $\dot{\eta}$ is the normalised contractile element velocity, given by the ratio between the lengthening velocity of the muscle ($\dot{\epsilon}$), and its maximal value ($\dot{\epsilon}_{max}$).

The force generated by the PEE is a function only of its length. An exponential relationship has been considered in this case [13, 56], with b_1 and b_2 muscle dependent constants:

$$F_{PEE} = b_1 \cdot e^{b_2\cdot\epsilon} - b_1 \tag{5.7}$$

The scalability of the muscular action is achieved by scaling the PCSA of the muscles with respect to the product of hand length and hand breadth parameters [21] from their values for $\overline{HL} = 18.22$ cm and $\overline{HB} = 8.00$ cm.

$$\frac{PCSA(HL, HB)}{PCSA(\overline{HL}, \overline{HB})} = 1 + 0.01333 \cdot (HB \cdot HL - \overline{HB} \cdot \overline{HL}) \tag{5.8}$$

Table 5.3 Muscles modelled on each skeletal chain

Index	Medial	Ring	Little	Thumb
1st FP	2nd FP	3rd FP	4th FP	APB
1st FS	2nd FS	3rd FS	4th FS	FBB
1st EDC+EI	2nd EDC	3rd EDC	EDQ	OPP
1st LU	2nd LU	3rd LU	4th LU	ADD
1st DI	2nd DI	4th DI	3rd VI	1st DI
1st VI	3rd DI	2nd VI	FDQ	APL
			ADQ	EPB
				FPL
				EPL

acronyms in the nomenclature section

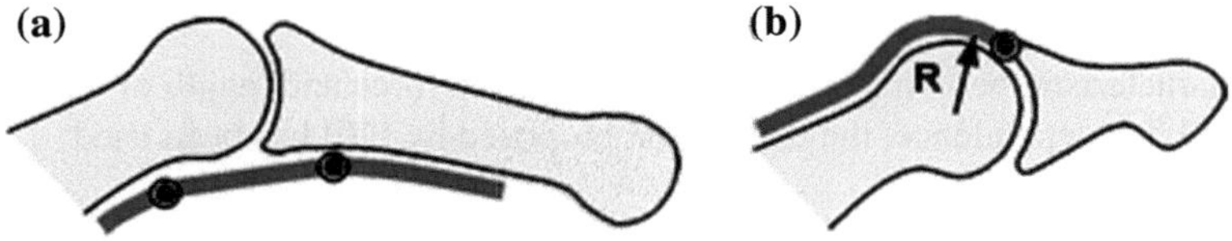

Fig. 5.8 Models for the tendons crossing the joints: (**a**) Straight lines, (**b**) Landsmeer's model I

The muscles considered on each skeletal chain are listed in Table 5.3. $PCSA$ data for index finger muscles have been taken from [58]. Data for the remaining muscles have been obtained from [27]. The muscle stress limit (S_{max}) has been obtained from [59]. Fibre and muscle lengths and the constants b_1, b_2 for index finger muscles have been taken from [13]; data for the remaining extrinsic muscles have been obtained from [60] and for the remaining intrinsic muscles from [61]. The muscle maximal lengthening velocity ($\dot{\epsilon}_{max}$) has been taken to be 2.5 s^{-1} [56].

Most of the muscles do not act directly on the bones, but transmit the force to the tendons, which finally insert into the bones. To model the tendon action crossing the joints, straight lines connecting 2 points have been considered, one fixed with respect to the proximal bone and the other one with respect to the distal bone (Fig. 5.8a). This approximation has been found to be close enough to the behaviour of all tendons with the exception of extensors [30], for which Landsmeer's model I has been considered (Fig. 5.8b). The data for the points defining the tendon actions have been obtained from [30].

The extensor hood mechanisms of the fingers are modelled as a tendon net. The net allows for the connection and division of the tendon paths. The insertions and connection points considered for the tendon nets on each skeletal chain are shown in Fig. 5.9. Appropriate force balances have been considered in the connecting points of this deformable tendon net. Second DI, fourth DI and ADQ tendons do present a double insertion into the proximal phalanges and into the extensor aponeuroses. A force distribution proportional to the amount of fibres of each branch [62] has been considered.

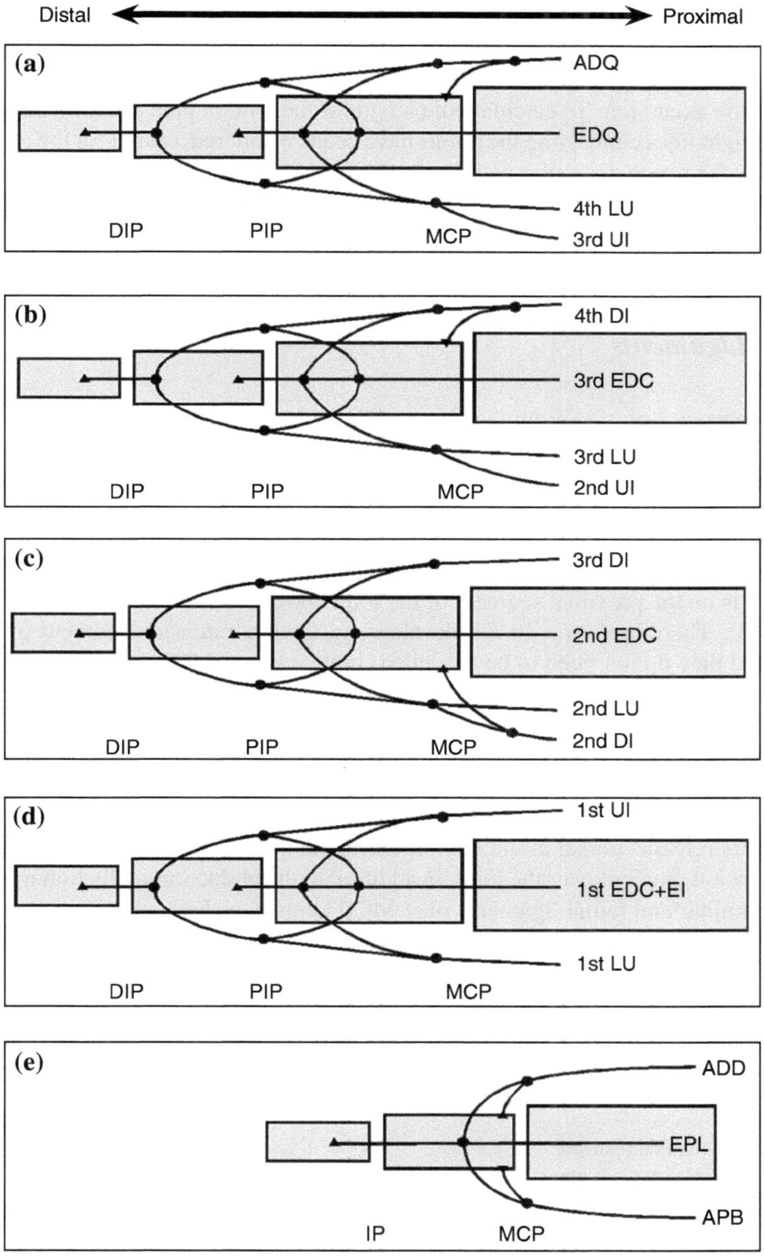

Fig. 5.9 Sketch of the extensor mechanisms of the fingers and thumb (dorsal view) showing the insertions into the bones (△) and the connections and splittings considered (∘): (**a**) little finger, (**b**) ring finger, (**c**) medial finger, (**d**) index finger, (**e**) thumb

The muscle force-length and force-velocity relationships presented above require the calculation of the lengthening of the muscles from l_o as a function of time. Having considered the tendons inextensible, the muscle lengthening coincides with the tendon excursion. To calculate the length of the tendon path crossing each joint (l_i), straight lines connecting the points have been considered, except for the extensor tendons, for which a circular path has been considered.

The data for the location of the points defining the tendon paths comes from [30, 47], and are also scaled with respect to the hand length and hand breadth [49].

5.5.3 Ligaments

The importance of modelling the effect of ligaments for studying free finger movements was proved in a previous work [18]. In the case of grasping, their consideration is not so relevant. Their effect can be neglected for studying power grasps, but they can play an important role in the case of some precision grasps, particularly those involving fast movements.

In the case of DIP and PIP joints of fingers and thumb, the insertion of the collateral ligaments on the proximal segment of the joint corresponds to the flexion-extension axis [63]. Therefore, they do not develop any flexion-extension moment over the joint and they do not need to be modelled. In the case of MCP joints, the proximal insertion of the lateral ligament on the metacarpal head remains dorsal to the center of the articular curvature (Fig. 5.10), so that collateral ligaments are lax in extension, but they become taut in flexion, decreasing significantly the range of lateral movement [46, 63, 64]. Tension on the radial and ulnar ligaments increases with adduction and abduction of the MCP joint, respectively. Furthermore, the line of action of the ligaments remains dorsal to the flexion-extension axis of the joint [64], developing an extension moment over the joint, in addition to the abduction-adduction moment.

Both ulnar and radial ligaments over MCP joints have been modelled. A unique fibre for each ligament has been considered, joining two points representing the insertions into the bones. One point is fixed with respect to the metacarpal, and the other one with respect to the proximal phalanx. No interaction between bone and ligament has been considered; therefore the ligament path is a straight line between the insertion points. Its non-linear behaviour has been taken into account considering

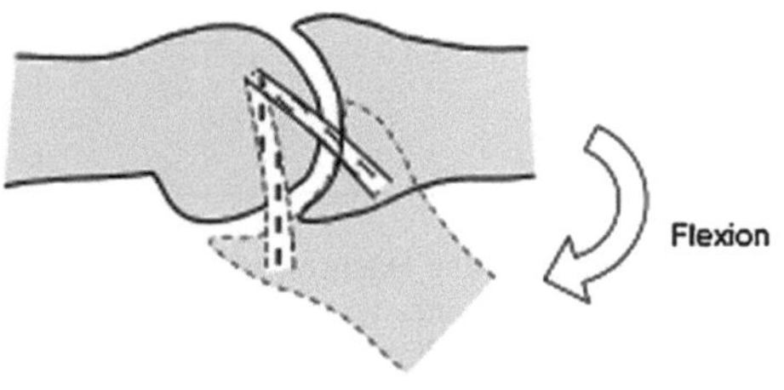

Fig. 5.10 Collateral ligament over MCP joints becomes taut with flexion

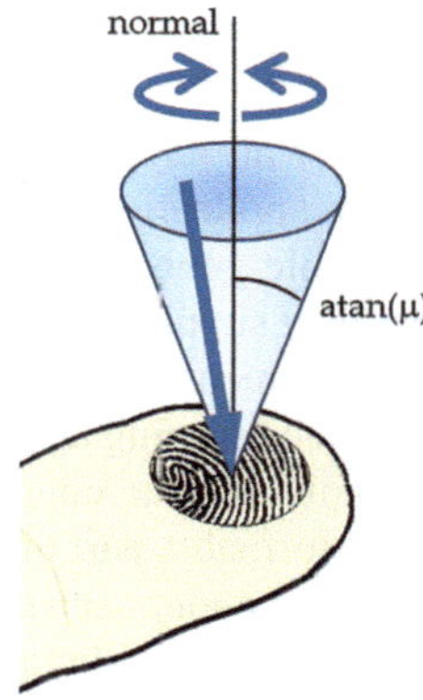

Fig. 5.11 Soft-finger contact model

a quadratic relationship between the force developed by the ligament (F_{lig}) and its elongation [65]:

$$F_{lig} = K \cdot (L_{lig} - L_{lig,o})^2 \tag{5.9}$$

where K is the characteristic constant of the ligament, L_{lig} the length of the fibre representing the ligament, and $L_{lig,o}$ the unstrained length of the ligament.

The data for the ligament insertion points have been obtained from the geometric model presented in [29], and the stiffness constant has been estimated to be 750 N/cm^2 from [66].

5.5.4 Soft Contact Model

The contact forces between the object and the hand have to be considered when dealing with the estimation of the muscle forces required for grasping an object. Unlike what happens with robots, real human fingers conform to the grasped object's shape. As the contact finger surface is deformable, the contact does not occur at just one point but over some finite area that increases as the normal forces increase. Due to this effect, in addition to the normal force and tangential force due to friction, human finger contact may support frictional torsional moments with respect to the normal at the contact point (see Fig. 5.11). This clearly shows that the consideration of rigid contacts, commonly used in robotics, is not appropriate for its use in studying the human grasp, and a soft contact model has to be used. Most objects manipulated by human hands are much stiffer than human hands and, therefore, it is reasonable for those cases to consider the grasped objects as rigid bodies and the hand as a deformable body.

Different approaches have been developed over the years to model soft contacts that fall into three categories: analytical elasticity-based models, elastic foundation models (EFM) and finite element models (FEM). A description of them can be found in Sect. 2.2.2.

They all have advantages and disadvantages, and have to be chosen depending on the application. Analytical models are the easiest to calculate but are restricted to simple geometries. FEMs have been increasingly used over recent years given that they supply information about the sub-surface stresses and strain in volumetric finite elements. However, they are excessively time consuming for fast simulation in dynamic grasping and manipulation models. EFMs were developed in order to allow a simple discrete contact calculation in more general surface geometries modelling the deformable part of the contact as a layer over a rigid base, and a series of discrete and independent spring in the contact normal direction. Therefore, we have chosen EFMs to model human contacts which allow us to use complex geometries but are not so time consuming. Several studies have used this type of model to study finger contacts [67–72].

In this work a soft contact model based on that of [67] was used. The friction constraints are derived based on general expressions for non-planar contacts of elastic bodies, taking into account the local geometry and structure of the objects in contact. The following approximation can be used to express the constraint relating the magnitudes of frictional force (f_t) and moment (τ_n):

$$f_t^2 + \frac{\tau_n^2}{e_n^2} \leq \mu^2 \cdot P^2 \tag{5.10}$$

where P is the total load applied in the direction of the contact normal, μ is the friction coefficient and e_n is called the eccentricity parameter [height of the ellipsoid described by Eq. (5.10)]. Considering a Winkler elastic foundation [73] of depth h and elastic modulus K, the eccentricity parameter is given by:

$$e_n = \frac{8}{15}\sqrt{a \cdot b} \tag{5.11}$$

where a and b can be calculated from the relative radii of curvature R' and R'' of the objects in contact and the compression δ of the elastic layer:

$$a = \sqrt{2 \cdot \delta \cdot R'};\, b = \sqrt{2 \cdot \delta \cdot R''};\, \delta = \sqrt{\frac{P \cdot h}{K \cdot \pi \cdot (R' \cdot R'')^{1/2}}} \tag{5.12}$$

The value of the human skin friction coefficient μ is difficult to measure consistently, because it is affected by many factors that not only depend on the skin itself (such as its texture, smoothness or hydration), but also on its interaction with external surfaces and the environment. In the literature, different methods of skin friction measurements have been proposed (see [74] for a review). In this work, we have used the value of $\mu = 0.8$ obtained from [75] as a first approximation. Additionally, the value for the stiffness of human fingertip K was assumed to be $K = 1{,}500\,\mathrm{N/m}$ obtained from [76].

5.5.5 Skin Model

The model has to incorporate a realistic geometry of the skin in order to provide the required contact information for evaluating grasps and to use it for different applications such as assessing the use of handheld products from an ergonomics point of view. The advances in computer animation have made possible the development of a number of convincing surface skin models.

Early models of the hand [77] were actually kinematic models that simulated roughly the external geometry of the hand and its movements. The geometry of the hand has been modelled mainly by jointed cylinders and cones [19, 20, 43]. This was the first approximation used in our model (see Fig. 5.12).

In order to conduct the collision detection in an efficient way, we have modified the geometry of the hand surface and the grasped object modelling them using the spherical extension of polytopes (s-topes). This graphical representation was successfully used previously in robotics [78], allowing a fast and efficient collision detection between the grasping hand and the grasped object while showing a sufficient level of realism (see Fig. 5.13). Collision detection was carried out by calculating the minimum distance between s-topes, based on the Gilbert-Johnson-Keerthi algorithm [79]. The algorithm was also used to calculate the minimum distance points that define the normal direction to the contact surface.

However, if the geometry of the hand model is not very accurate, the algorithms for inverse kinematics are not precise enough. Recently, some efforts have been made in accurately modelling the surface of real hands to be incorporated into 3D hand models. [80] presented an automated method to make a specific human hand model from an image of the palm of the hand. Different algorithms were used in the process: principal creases are extracted, joint locations are estimated from them and the skin geometry of a generic hand model deformed based on hand contours. [81] made a scalable 3-D geometric model of the hand based on 66 landmarks of

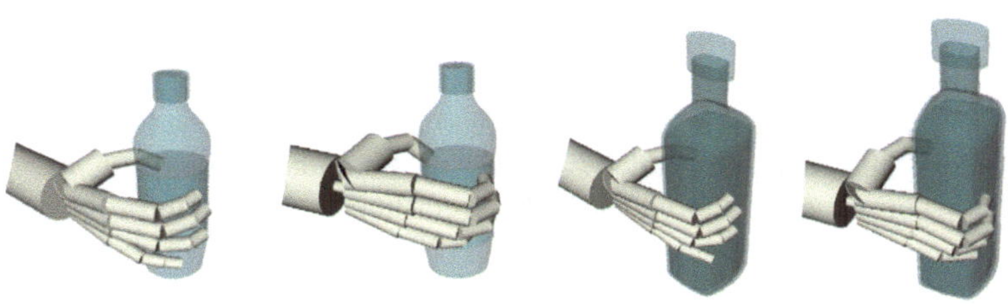

Fig. 5.12 Different views of the geometric model using cylinders to simulate two grasps of different bottles

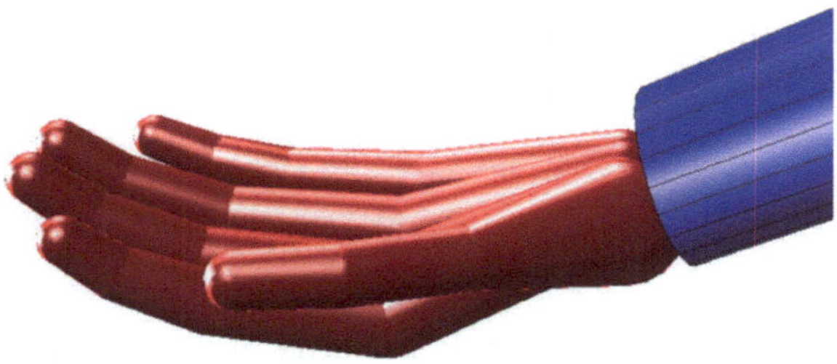

Fig. 5.13 External geometrical representation of the hand with s-topes

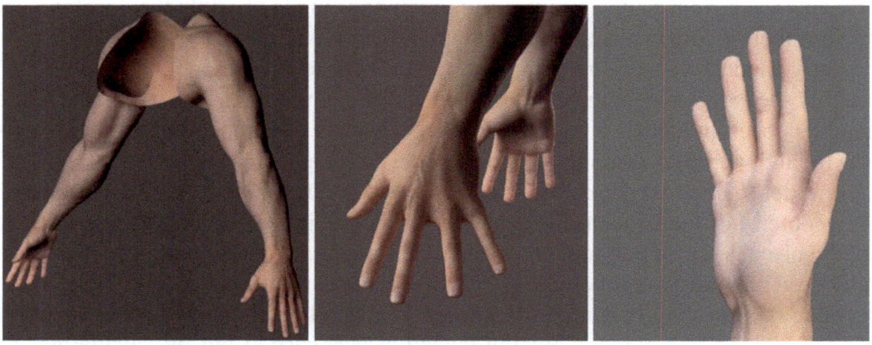

Fig. 5.14 Acquired 3D polygonal mesh for the surface of the hands and arms

the palm surface from 100 subjects in four functional postures. The purpose was to analyse the deformation of the palm surface during the grasp of an object. Recent models incorporate the surface of the hand as a mesh object with more or less realism, obtained from the location of a number of landmarks of the hand or from digital 3D-scanning of the hand [1, 35, 82]. The mesh is linked to a skeleton whose movement controls the deformation of the mesh with different types of algorithms.

Therefore, in order to create a more realistic model of the hand surface, we propose to use a surface skin model similar to that of [1] or [3]. We have acquired a surface skin model consisting of a 3-dimensional polygonal mesh for the surface of the two human hands and arms (see Fig. 5.14). In order to make the model scalable, we have adjusted each of the segments accordingly with the standard ergonomic measurements ($HB = 87\,mm$, $HL = 189\,mm$) corresponding to the 50th percentile.

The geometry of the skin model is defined at only one opened posture. It gets scaled when the kinematic model is scaled to the anthropometry of the specific subjects or population group under study, according to the length of the segments defined by the scalable biomechanical model [20, 49]. Figure 5.15 shows the model scaled to different percentiles of population by gender.

The model is currently composed of rigid parts which are not deformed with the movement of joints. This is enough for the current applications where only the

	Male			Female		
	5%	50%	95%	5%	50%	95%
HB	79	87	95	69	76	83
HL	173	189	205	159	174	189

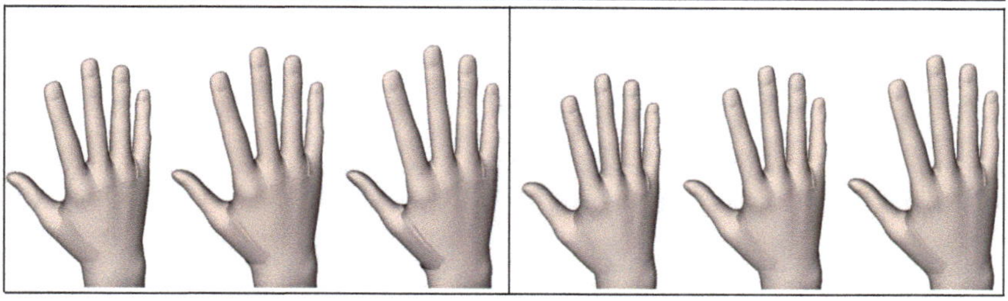

Fig. 5.15 Scaled skin model to different subjects anthropometry according with standard percentiles of population by gender

fingertips contact the object. For more complex grasps, where the contact zones correspond to the joint between segments, a skin deformation algorithm should be implemented. This skin algorithm will also improve the visualization of the grasp giving a more realistic hand posture. It should define the deformed geometry of the surface skin model when the posture of the kinematic model is changed. The algorithm assigns each bone a capsule-shaped envelope. Vertices of the modified skin within these envelopes move with the bones. Where envelopes overlap, vertex motion is a blend between the envelopes. The influence of each bone for vertices within the intersection of two bones' envelopes is controlled by assigning weight values. The ratio of a vertex's weight values, which always total 1.0, determine the relative extent to which each bone's motion affects the vertex.

5.5.6 Closure Algorithm

As stated before, the model has to simulate and show the grasping of an object in a realistic way. To satisfy this requirement, it is not enough to have a visually realistic model of the surface skin. The model must also be able to predict feasible grasping postures.

The grasping posture algorithm is based on the calculation of appropriate rotation rates for the joint angles from the use of two characteristic hand postures: the most open posture (MOP) and the tentative grasping posture (TGP) shown in Fig. 5.16. For the experiments performed for this book, these postures have been experimentally measured by means of data gloves or motion capture systems. However, the experiments are tedious and time-consuming as they have to be carried out for every subject, task, and object. For that reason, there are no hand models valid for the study of general grasping. As future work, in order to overcome this drawback and

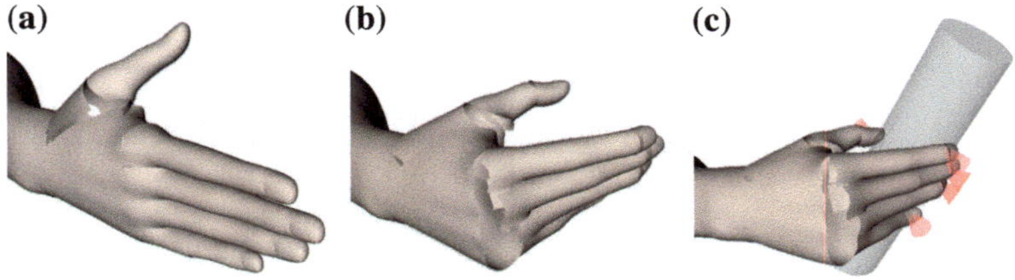

Fig. 5.16 Characteristic hand postures: (**a**) Most open posture (MOP), (**b**) Tentative grasping posture (TGP) and (**c**) Final grasping posture (FGP)

to automatically generate the characteristic postures MOP and TGP, artificial neural networks could be used as proposed by [83].

When trying to simulate the grasping of an object with the biomechanical model of the hand, we can't directly use the grasping postures measured experimentally, as they do not exactly match the simulated grasped object, becoming a non-conforming grasp, so that contact information (contact points and normals) cannot be obtained). To avoid this, we propose to use a closure algorithm based on that of [84]. This algorithm uses a function to automatically generate a natural grasping path of the hand model from a fully opened state to a clenched one. The goal is to find contacts between the surface hand skin and the object surface while rotating the joint angles of the fingers. It is very important to choose appropriate rotation rates for the finger joints, as they affect the final posture prediction [43]. We have solved this situation by using rotation rates that try to match those experimentally observed. The rotation rates are defined by the difference between the angles of the most open posture (MOP) observed when approaching the hand to grasp the object, and the clenched one once the grasp is performed, which will be used as a tentative grasping posture (TGP).

A geometric model of the hand, together with a contact model are required to generate the grasp because, at each rotation step, contact has to be checked between the surface skin model and the surface of the object model. In real grasping, the surface of a hand deforms in a non-linear way when making contact with the object. To avoid long execution times, we considered a geometric collision-detection algorithm. The hand segments are considered as rigid bodies, and their deformation is simulated allowing the penetration of the surface skin model and the object model. This penetration is limited by a tolerance that relates to the hand stiffness at each contact region. A maximum penetration of 3 mm has been considered for all hand segments as a first approximation.

The distances between the points on the skin surface and the object are calculated while each joint rotates according to the specific joint rotation algorithm. When this distance reaches the given maximum penetration tolerance for a given segment, the contact of that segment is achieved and the joint rotation ends, together with the rotation of all proximal joints. When the distal segments of all four fingers make contact with the object, the grasping simulation terminates. A summary of the steps taken to simulate the closure of the hand is outlined in Algorithm 1.

Algorithm 1: Hand closure algorithm

```
input  : most open posture (MOP) and tentative grasping posture (TGP)
output: final grasping posture (FGP) and contact information
begin
    set maximum penetration = 3 mm;
    foreach joint of the hand do
        calculate rotation rate to reach from the MOP to the TGP
    foreach finger do set flag to zero;
    while all finger flags ≠ 1 do
        foreach joint do increase joint angle according to the calculated rate;
        if angle exceeds the joint limits then angle = limit;
        foreach phalange do
            penetration = 0;
            get collision information;
            if collision is true and penetration > maximum penetration then
                get contact point and normal direction;
                return finger flag = 1;
    return final grasp posture and contact information for each finger;
```

5.5.7 Neuromuscular Control

The actual grasping forces for a given posture will be obtained by considering that they have to satisfy the dynamic equilibrium of the hand and the grasped object. The movement of the skeletal chains, together with the contact forces and the corresponding application points are input to the model. The problem to be solved is the derivation of the muscle activation levels required to produce the given motion under the external loads. It is, therefore, an inverse dynamics problem.

The dynamics equations of the open chain of rigid bodies have been derived using the Lagrange method [85]. For a system with m generalised coordinates q_k, this equation is expressed as:

$$\frac{d}{dt}\frac{\partial L}{\partial \dot{q}_k} - \frac{\partial L}{\partial q_k} = Q_k^{nc} \qquad k = 1, \ldots, m \tag{5.13}$$

where L is the Lagrangian function and Q_k^{nc} are the generalised non-conservative forces. The generalised coordinates have been considered coincident with the system DOF (m = 23).

Equation (5.13) together with the force balances of the tendon nets makes up the equilibrium equations of the grasping hand (49 equations). The equilibrium of the grasped object is defined by six more equations. A total of 55 equations with 99 unknowns (muscle and tendon forces and contact forces and moments) form the final grasping mathematical problem, along with the inequalities given by the muscle model (lower and upper bounds of muscle forces and lower bounds of tendon forces) and the soft contact model (one inequality by contact point). There is not a unique combination of muscular efforts that satisfy the equilibrium constraints.

To solve the problem, a criterion chosen by the central nervous system CNS to determine the muscle action control must be introduced. The most commonly used criterion in the literature is the maximisation of the endurance [31], through the minimisation of the non-linear objective function

$$OBJ = \sum_{i=1}^{l} \left(\frac{F_i}{PCSA_i} \right)^n \tag{5.14}$$

with n between 2.0 and 4.0 (2.0 being the most used), and where l is the number of muscles (34 in our model), F_i represents the force exerted by muscle i, and $PCSA_i$ its physiological cross-sectional area. This function is minimised when subjected to Eq. (5.12) together with the force balances of the tendon nets. Additional constraints are that tendon forces must be non-negative, and the limits of muscle forces obtained from Eqs. (5.2) and (5.3) varying the muscle activation level from 0 to 1

$$F_{PEE} \cdot F_{max} \leq F \leq (F_l \cdot F_v + F_{PEE}) \cdot F_{max} \tag{5.15}$$

The validity of using the maximisation of fatigue resistance (5.17) to solve the indeterminate problem for grasping simulation has not been proven, therefore in Sect. 5.7 experiments performed to test the accuracy of this function and the proposal of others are detailed.

5.6 Simulation Framework for Human Hand Grasping

There are many elements of the musculoskeletal system that interact to enable coordinated movement and object grasping, as was shown in the previous section. The challenge is to develop synthesizing detailed description of these elements to create an integrated understanding of human grasping and to establish a scientific basis for improving manipulation skills of artificial hands or correcting abnormal movement.

In order to accomplish this, a framework, in combination with experiments, is needed. It must reveal the cause-effect relationships between neuromuscular excitation patterns, muscle forces, contact information and quality of grasp. Dynamic simulation of movement can provide such a framework, integrating models describing the anatomy and physiology of the elements of the neuromusculoskeletal system and the mechanics of multi-joint movement. Muscle-driven dynamic simulations complement experimental approaches by providing estimates of important variables which are difficult to measure experimentally and enables cause-effect relationships to be identified.

The inability to reproduce results is a major limitation to advancing the science of biomechanical simulation. Many laboratories develop their own software and do not make it free or open-source. Researchers typically must spend a great deal of time implementing each new simulation and creating tools to analyse it.

In this work, the biomechanical model described in the previous section has been implemented using different software tools that have been chosen for their versatility, flexibility and open access. The following sections give an overview of the system and details of each part are presented.

5.6.1 Related Work

Different software frameworks have been proposed in the literature to model the human hand [86] presented a skeletal musculo-tendon model for the human hand and forearm, called "HelpingHand" (see Fig. 5.17b). It enables the simulation of forward dynamics predicting finger position given a set a muscle activation. They also proposed a solution for the inverse problem of determining a set of muscle activations to achieve a given pose. This work is based on the model proposed by [87] which, besides using the conventional modeling of muscles to control the movement of the hand, it converts these contraction values into deformation of the skin (see Fig. 5.17a). Although these models are anatomically realistic, their implementation have been done using *Maya*, a commercial 3D animation platform, and the authors have not made it available, so it will be necessary to re-implement the parts deemed useful.

A very popular musculoskeletal modeling environment, called SIMM (Software for Interactive Musculoskeletal Modeling),[1] was introduced in the early 1990s by [88, 89]. Using SIMM, anatomically realistic musculoskeletal models of the lower and upper extremities can be created for different applications such as examining the biomechanical consequences of surgical procedures. Figure 5.18 present models of different body parts developed by [90–92] using SIMM. Especially relevant is the 3D kinematic model of the upper-extremity (from the shoulder to fingertip) developed by [92] in SIMM, with dynamic simulations performed using SIMM/Dynamics Pipeline/SDFast. Although it has been widely used, SIMM is not freely available, doesn't provide assistance with the computation of muscle excitations and has limited tools for analyzing the results of dynamic simulation. Additionally, it doesn't provide a way to model the skin and its contacts with objects which are indispensable for grasp simulation.

Over the past decade, new software engineering methods have emerged that enable the development of software systems that are more extensible [93] created an

[1] http://www.musculographics.com/

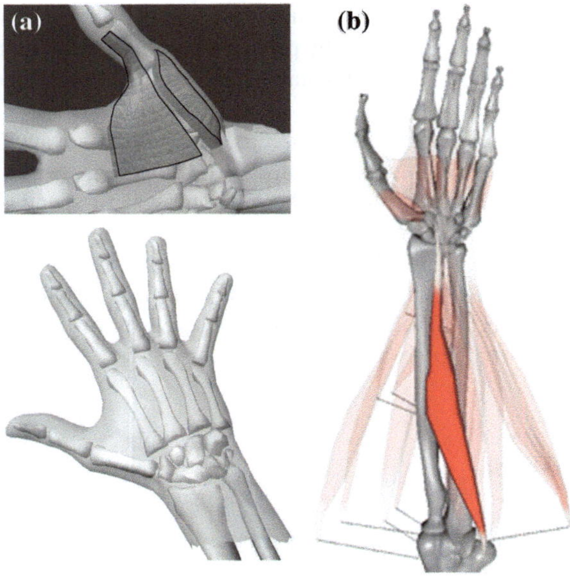

Fig. 5.17 Existing musculoskeletal human hand models. (**a**) Skin and bones by [87]. (**b**) Helping-Hand by [86]

open-source software system called *OpenSim*[2] which lets users develop models of musculoskeletal structures and create dynamic simulations of a variety of movements. It has been widely used specially for studying the movement of lower extremities, although some attempts have also been made to study the upper extremity movements. Previously developed kinematic models of the upper-extremity using SIMM [90, 92] were ported to OpenSim and are now available Fig. 5.19.

OpenSim is being used by several laboratories over the world as a software framework for biomechanics simulation given that it is freely available. The user can extend it by developing customized controllers, analyses, contact models, and muscle models. However, as with the previously mentioned software frameworks, it focuses on modeling the musculo-skeletal system of the hand, but it lacks some features to simulate object manipulation. For example, it allows one to incorporate meshes to analyse contacts but it is not simple to scale them to model the different hand sizes, include algorithms to close the hand or add virtual objects and extract contact information when interacting with them. As future work, it could be interesting to study how our biomechanical model can be implemented using OpenSim and to develop the functions that OpenSim currently lacks.

GraspIt![3] is a robot simulator developed for robot grasping by [94] that includes a model of the human hand (see Fig. 5.20). Given that it has been designed for grasp

[2] https://simtk.org/home/opensim

[3] http://www.cs.columbia.edu/~cmatei/graspit/

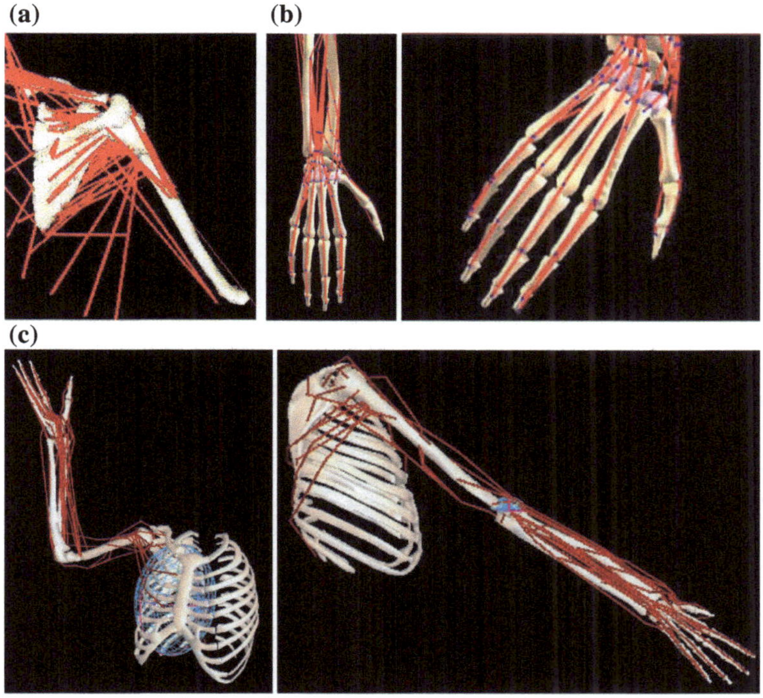

Fig. 5.18 Musculoskeletal models using SIMM (Figures taken from the SIMM Gallery). (a) Shoulder [91]. (b) Wrist [90]. (c) Upper extremity model [92]

Fig. 5.19 Musculoskeletal models of the human hand using OpenSim

analysis, many of the same capabilities used for robots can be adapted to study the human grasp. Unlike the previous frameworks, it comes with a model of the human hand skin that can be used to detect contacts with the virtual objects when grasping them. Additionally, it has grasp quality measures implemented to evaluate a given set of grasps. The skin is implemented as a set of rigid bodies that does not deform when bending the hand joints. However, as mentioned in the first part of this manuscript,

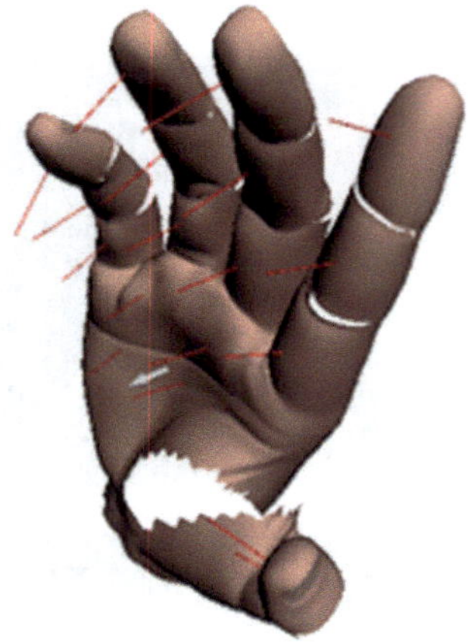

Fig. 5.20 Human hand model available in GraspIt!

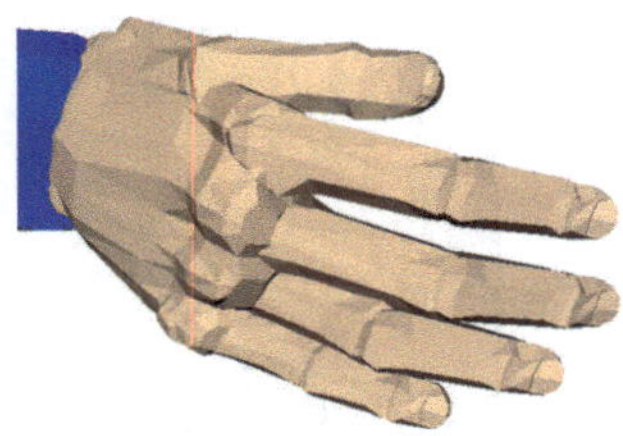

Fig. 5.21 Human hand model available in OpenRAVE

the software framework itself has a rather monolithic and less modular architecture that makes it difficult to improve, add functionality and integrate with other tools and frameworks. In addition, it does not provide a convenient Application Programming Interface (API), which allows script programming.

OpenRAVE[4] is another open-source robotic simulation environment developed by [95] that has also been widely used by the robotics community and which has also been used for our robot grasping simulations in the first part of this work. As mentioned above, it has been designed as an open architecture targeting a simple integration of simulation, visualization, planning, scripting and control of robot systems. It has been written in C++ and has a modular design, which allows its extension and further development by other users. It comes with a model of a human which has a simple model of the hand shown in Fig. 5.21 but which had to be adapted to the kinematics and scalable capabilities of our model.

Given that none of the available software frameworks meet all our requirements, we decided to implement our biomechanical hand model using OpenRAVE given our previous experience with it for robot grasp simulation. OpenRAVE has the capability to communicate with MATLAB, which enabled us to reuse the existing code made by [49] for the neuromuscular control and create separate modules in MATLAB,

[4] http://www.openrave.org/docs/latest_stable/

C++ and python to model the remaining parts. In the following section an overview of the complete system is presented.

5.6.2 System Overview

We have developed a framework for modelling, simulating and analysing the grasp performed by human hands. It includes low-level computational tools that are invoked using different applications. The architecture of the system is shown in Fig. 5.22.

At the lower level we have chosen OpenRAVE for kinematic definition and visualization. OpenRAVE is designed as a plugin-based architecture where a plugin offers implementations of base interface classes that are loaded dynamically into the environment. For our system, several plugins available in OpenRAVE are used, specially the collision engine developed to interface with ODE (Open Dynamics Engine) which plays a major roll giving contact information when performing grasping simulation. Another plugin that can be optionally used is a viewer that provides 3D visualization of the environment. At the moment, Coin 3D/Qt is used for this purpose.

A new plugin for OpenRAVE called *OpenHand* has been developed as part of this work. It is a plugin extending the OpenRAVE *module* interface to provide the simulation of the hand biomechanical model presented in the previous section. It has

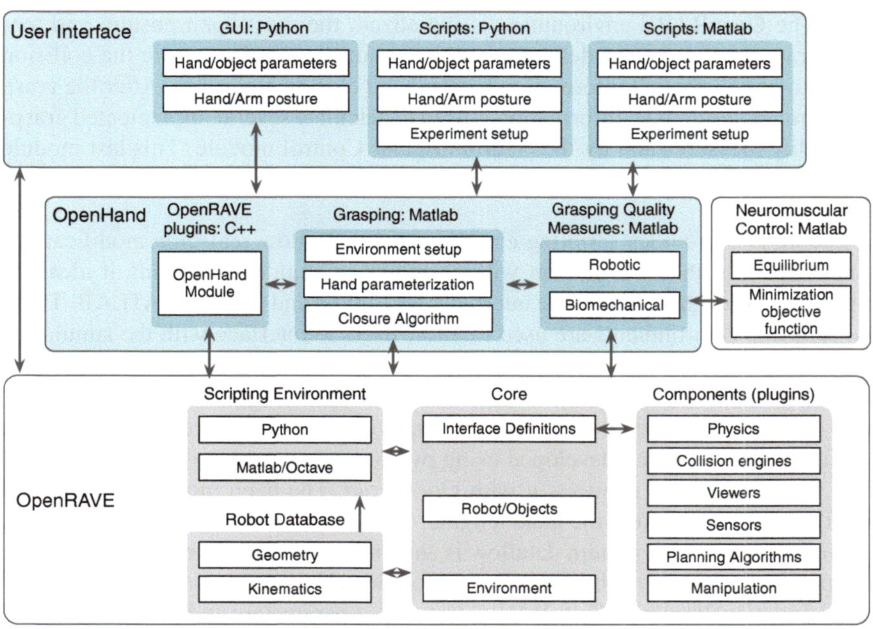

Fig. 5.22 Architecture of the system: (*green*) new modules interfacing with (*grey*) existing software

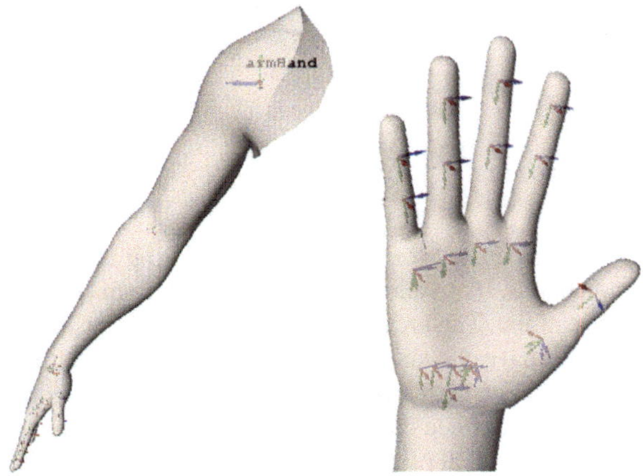

Fig. 5.23 Arm and hand models created by the OpenHand plugin showing its joint kinematic axes

the definition of the kinematic model enabling users to scale it with the hand breadth and the hand length anthropometric parameters (see Fig. 5.23). In order to control the kinematic model, different functions are provided to specify a given posture using the angles of the 23 DOF of the hand and 7 DOF of the arm.

Each grasping experiment is defined through a Grasping module using MATLAB. The OpenRAVE environment is initialized, the most open posture and tentative grasping posture are defined, then the simulator interfaces with the collision engine to obtain contact information for the hand closing algorithm. After the grasp is performed, the contact information is used to calculate several implemented grasping quality measures and by the Neuromuscular Control module. This last module was previously developed using MATLAB and its optimization toolbox (version R2008b) [49].

OpenRAVE provides scripting environments that allow real-time modifications to any aspect of the environment without requiring shutdown, making it ideal for testing new algorithms. It provide interfaces for Python and Octave/MATLAB. These two scripting environments are used to allow users to interface with the simulation model to define hand and object parameters, setup the grasp posture and run several experiments.

In order to move each joint of the hand and visualise its posture, a graphical user interface (GUI) has been developed using python/Qt and it is shown in Fig. 5.24. It starts OpenRAVE and connects it with *OpenHand*. The hand model can be scaled with the hand breadth and the hand length.

An overview of the system dataflow is shown in Fig. 5.25. The hand parameters are specified as input to the *OpenHand* module which scales the hand kinematic model and visualizes it in OpenRAVE. The object parameters and position are used for the grasping module to set up the environment in OpenRAVE. The *most open* and *tentative grasp postures* of the hand are inputs to the *Closure algorithm* module

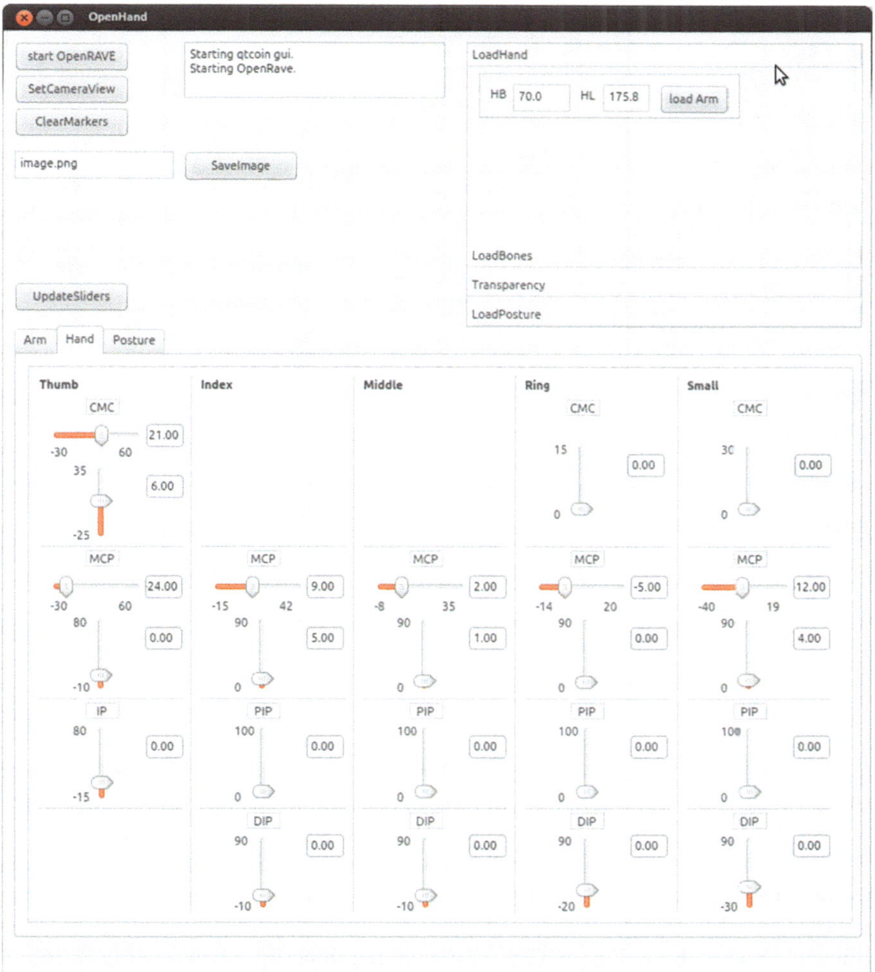

Fig. 5.24 Interface for apply forward kinematics on the hand model

and are currently obtained experimentally. These postures are used to determine the rates to move each hand joint until the fingers are in contact with the grasped object (see Sect. 5.5.6).

The contact points and normals as well as the final grasping posture are used by the *Neuromuscular Control* module to predict both the muscle forces and the contact forces and torques. The contact information is also used to calculate all implemented measures to assess the grasp quality.

We have used two sensor systems to capture the posture of the hand experimentally: an instrumented glove and a motion tracking system. Instrumented gloves have been widely used, with the Cyberglove® system being the most popular one (see

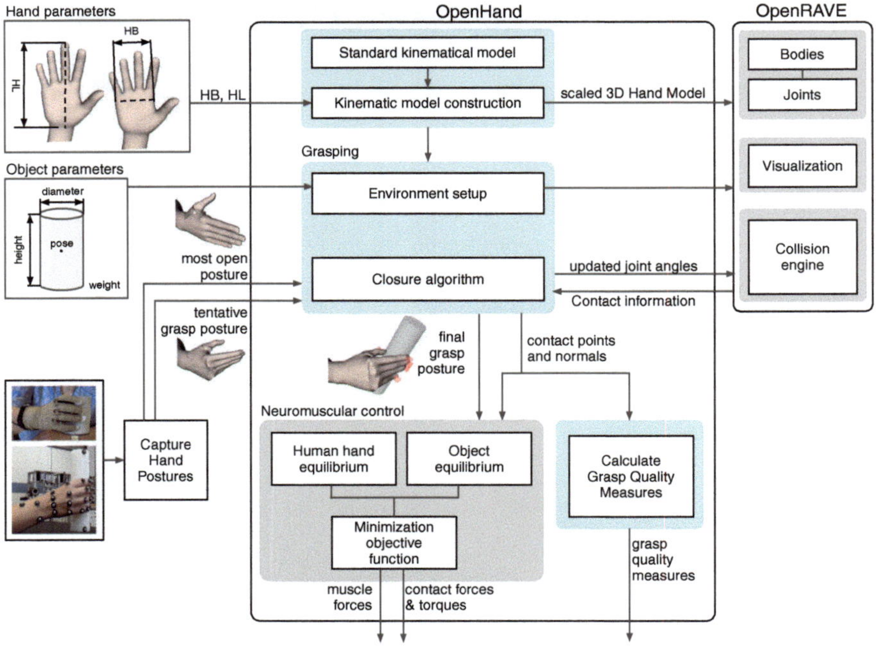

Fig. 5.25 Data flow diagram of the implemented system

Fig. 5.26a). Using this glove, we instrumented different subjects' hands to register hand posture data. The system was appropriately calibrated using the protocol created by [96]. This glove provides the main joint angles of the hand with an acquisition rate of 15 Hz. It has been proven to be valid for the measurement of hand postures during grasping tasks, with its sensors' repeatability errors ranging from 1.2° to 5°. When it was required to register finger force data, we have additionally instrumented the subjects' hand with the Finger TPS' system at the fingertips shown in Fig. 5.26b. It utilizes highly sensitive capacitive-based pressure sensors to reliably quantify forces applied by the human hand.

However, it is preferable to measure the movement by using techniques that do not interfere with the normal development of activities to be performed by the hand. In this regard, techniques of motion capture from video images that use passive markers (or videogrammetry) are good alternatives. Such visual techniques are based on the reconstruction of the coordinates of markers from images, placed on the tracked object. This is the technique used to obtain the majority of grasping postures used for the experiments of this work. The technique involves: placement of markers on the skin, registration of reference positions, record the movements, obtaining 3D coordinates of the markers, obtaining the 3D orientation of the different segments, and calculation of physiological rotation angles. A motion tracking system from Vicon consisting of 8 Vicon cameras Bonita Giganet was used (see Fig. 5.26c).

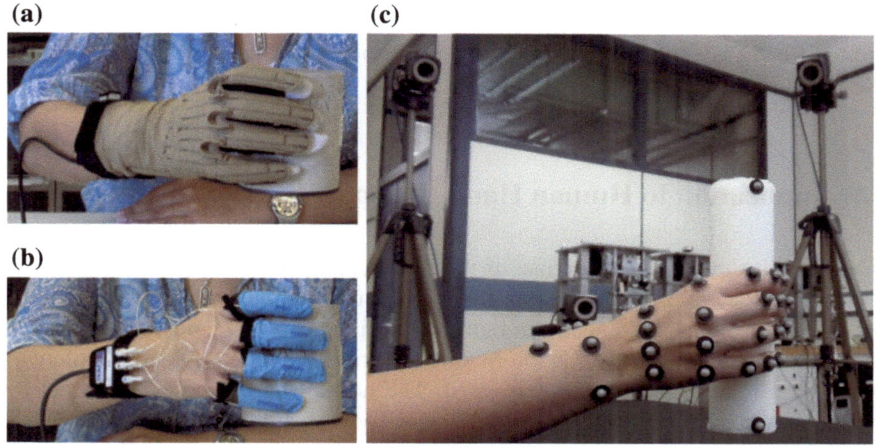

Fig. 5.26 Hand instrumented for posture data registration: (**a**) Cyberglove. (**b**) Finger TPS and (**c**) Vicon system

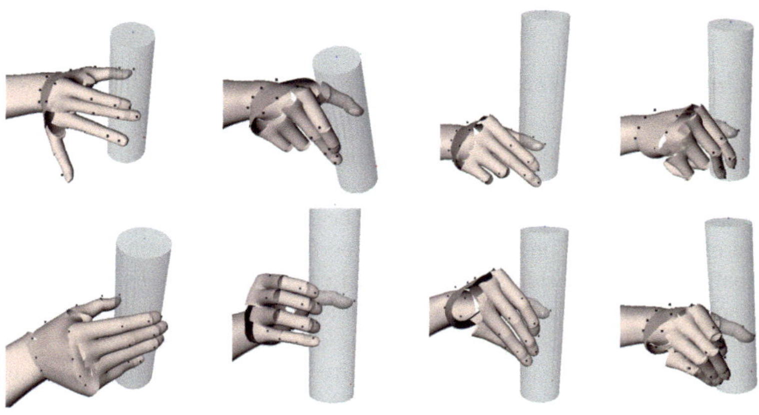

Fig. 5.27 Hand model showing different tentative grasp postures estimated with the Vicon system (position of marker reflectors showing as black points)

The hand postures have been obtained based on the technique recently developed by [97] which reconstructs the 3D coordinates of markers reflectors placed on the back of the hand recorded by videogrammetry. The coordinates of these markers are obtained in the reference postures and in the postures to be measured and are used to define a coordinate system on each segment in which the rotation angles of the joints are calculated. Examples of various tentative grasp postures of the hand grasping different cylinders estimated with the Vicon system are shown in Fig. 5.27.

An example of an initial environment showing different objects with the human arm and hand is shown in Fig. 5.28. Final grasp postures are estimated using the presented grasping framework.

5.7 Application to Human Hand Grasping Simulation

In this section, we present the use of the biomechanical model for simulating the human grasp by taking into account the equilibrium not only of the grasping hand but also of the grasped object, through the consideration of a soft contact model. In particular, the objective function required to solve the indeterminate problem of finding the muscle forces is investigated. The normal finger forces estimated by the model were compared to those experimentally measured. The popular objective function sum of the squared muscle stresses was shown not suitable for the grasping simulation, requiring at least being complemented by task-dependent grasp quality measures.

5.7.1 Material and Methods

The *Neuromuscular control* of the biomechanical model requires the posture of the hand, the contact points and the contact forces as input data (see Sect. 5.5.7). With these data, the model allows the estimation of the muscle forces required to counteract the given external forces on the hand while performing the given movements. For such a goal, the model considers the popular criterion of maximising the endurance in order to solve the indeterminate problem of finding the muscle forces. The validity of the model will be checked by reproducing a grasping experiment described in the following section. In particular, the validity of the criterion of maximising the endurance to solve the indeterminate problem is investigated.

5.7.2 Problem Solving and Neuromuscular Control

The problem to be solved is to find the muscle forces required to grasp the object. That entails to account for the equilibrium of the grasping hand and the grasped object. It is an inverse dynamics problem.

The equilibrium equations of the grasping hand (49 equations) is defined using Eq. (5.13) together with the force balances of the tendon nets. The equilibrium of the grasped object is defined by six more equations. A total of 55 equations with 99 unknowns (muscle and tendon forces and contact forces and moments) form the final grasping mathematical problem, along with the inequalities given by the muscle model (lower and upper bounds of muscle forces and lower bounds of tendon forces)

(a)
(b)

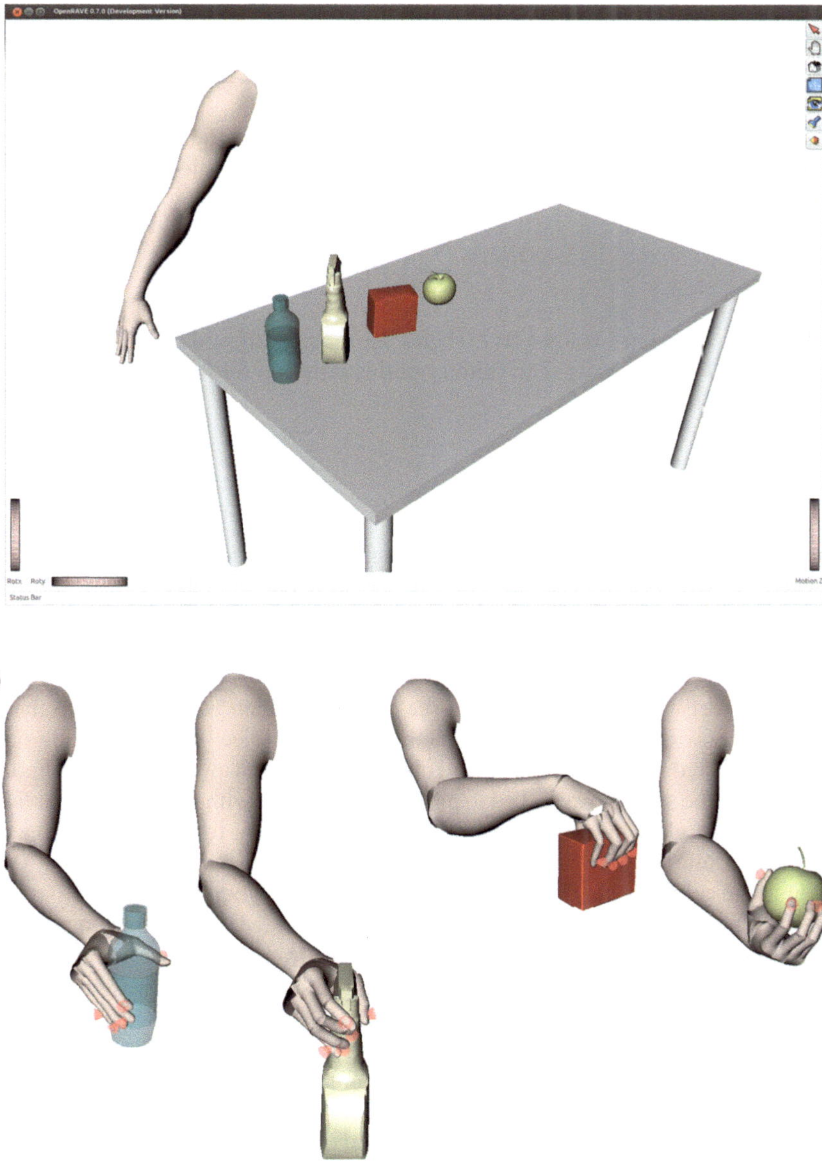

Fig. 5.28 Hand model showing different final grasp postures estimated with the grasping framework. (**a**) Initial environment. (**b**) Examples of final grasp postures (*red cones* represent the contact points)

and the soft contact model (one inequality by contact point). There is not a unique combination of muscular efforts that may satisfy the equilibrium constraints. To solve the problem, a criterion chosen by the CNS to determine the muscle action control must be introduced.

The most commonly used criterion in the literature is the maximisation of the endurance [31], through the minimisation of the nonlinear objective function

$$OBJ = \sum_{i=1}^{l} \left(\frac{F_i}{PCSA_i} \right)^n \tag{5.16}$$

with n being between 2.0 and 4.0 (2.0 being the most used), and l being the number of muscles (34 in our model). The validity of this criterion for the grasping simulation is checked in this study.

5.7.2.1 Validation Experiment

The validity of the model was analysed through the simulation of grasping cylindrical objects. An experiment was designed in which a female subject (age 32, height 1.61 m, weight 68 kg, HB 71 mm, HL 163 mm), appropriately instrumented, was asked to grasp alternatively two cylinders of different size and weight and hold them with their axes in vertical orientation (gravity direction).

The subject was seated by a table, the height of which was adjusted so that the subject's elbow coincided with the table height. The subject's arm was lying on the table in a relaxed posture, with the hand placed about 15 cm away from the cylinder to be grasped. The subject was asked to grasp each cylinder with her fingertips and hold it at a fixed height while keeping it in vertical orientation, for 2–3 s approximately, and then to return it to its initial location.

First, the subject's hand was instrumented with the Cyberglove system to register hand posture data. The system was appropriately calibrated [96]. The subject repeated the action three times for training without data registration, and five more times with posture data registration (Fig. 5.29a). Second, the subject's hand was instrumented with the Finger TPS system (Pressure Profile Systems, Inc.) at her fingertips to register finger force data. After the calibration of the system was carried out, the subject repeated the action three times for training without data registration, and five more times with force data registration (Fig. 5.29b). This procedure was carried out twice: first for a cylinder of 0.401 kg weight and of 64 mm diameter (cylinder 1) and second for a cylinder of 0.04 kg weight and of 82 mm diameter (cylinder 2).

The model was used to simulate the grasping of both cylinders. The simulation only considered the static case of holding the cylinders at a fixed height. To carry out the simulation, the subject's hand data (HL and HB) and the object data (weight and diameter) were input to the model, along with the most open posture (MOP) of the hand and the tentative grasping posture (TGP) registered with the Cyberglove. So far, the registered postures are required to generate the rotation angle rates that

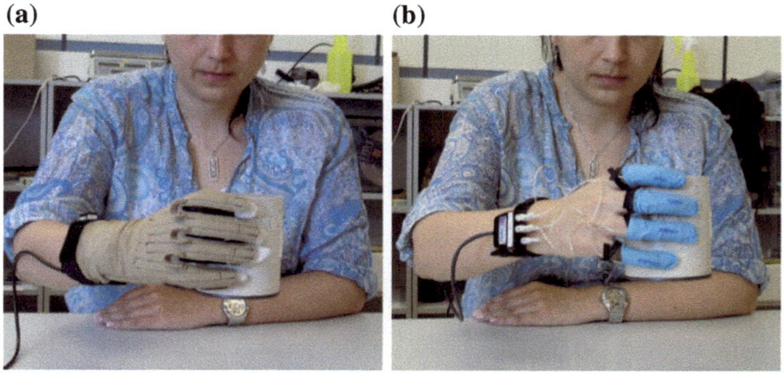

Fig. 5.29 The subject is holding the lighter cylinder (cylinder 1) of the experiment. (**a**) Hand instrumented for posture data registration. (**b**) Hand instrumented for finger force data registration

are used by the grasping posture generation algorithm to obtain the predicted final grasping posture, the contact points and contact normal directions. This information is used by the optimisation algorithm to obtain the contact forces and the muscle forces required for the equilibrium of the grasped cylinder and the grasping hand. In the future, this will be replaced by a neural network able to automatically obtain both postures and avoid the use of registered data as shown by [83]. The data reported by [75] were used to select the appropriate friction coefficient (0.8) between the hand and the cylinder material being grasped.

The results of the simulation of grasping both cylinders were the grasping postures, the contact points, the contact normal directions, the contact finger forces and torques, and the muscle force distribution. The normal finger forces estimated by the model were compared with those registered with the Finger TPS system. To investigate how the model could be improved to achieve better results, the simulation of grasping both cylinders was carried out under four different modifications of the model, described in the following section.

5.7.3 Results and Discussion

The hand movement pattern during the experiment can be observed in Fig. 5.30. This figure shows the joint angles registered by the Cyberglove system in one of the repetitions for cylinders 1 and 2. The hand starts from a relaxed posture. Just before grasping the cylinder, the hand gets open, which is seen mainly as an extension and abduction of MCP joints. The grasping is then achieved basically by means of the flexion of the different joints. Once the object is grasped, the joint angles registered during the static hold of the cylinders remain quite constant.

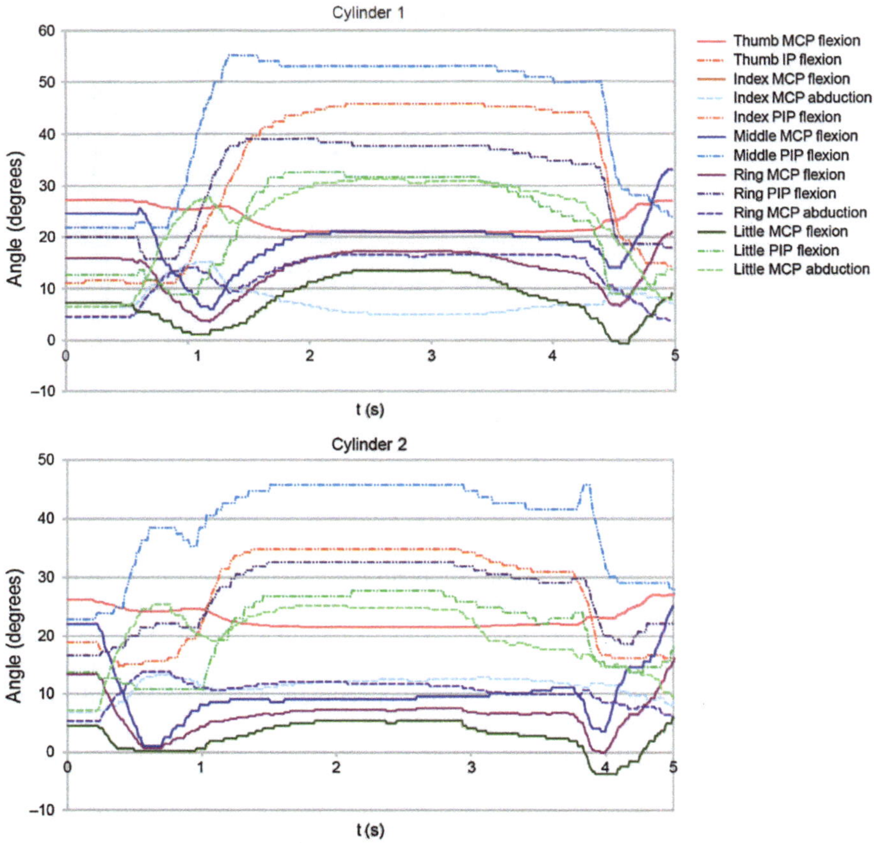

Fig. 5.30 Joint angles (in degrees) registered during one of the repetitions for grasping both cylinders 1 and 2

For each cylinder, the model needs the hand in the most fully open posture and in the hand grasping posture (as tentative) to calculate the joint rotation rates. The most fully open postures (Table 5.4) were obtained as the mean of the MOPs identified at each of the repetitions, being the standard deviation (SD) of the joint angles lower than 8.5.

For each repetition, the mean postures during the static hold of the cylinder were also obtained. The mean of these values for each cylinder was used to define the tentative grasping posture required to calculate the joint rotation rates (Table 5.5). Again, the SD of the joint angles among repetitions was lower than 8.5, which indicates that the experiment was repeatable. This makes it possible to interrelate the posture data and the force data registered in different repetitions.

Table 5.6 presents the joint angles calculated by the model (from the use of the collision-detection algorithm) for the grasping postures of cylinders 1 and 2,

Table 5.4 Mean (SD) joint angles defining the MOPs for grasping both cylinders 1 and 2

	MCC		MCP		PIP	DIP
	Flexion (°)	Abduction (°)	Flexion (°)	Abduction (°)	Flexion (°)	Flexion (°)
Cylinder 1						
Thumb	−8.0 (2.3)	49.0 (5.1)	25.1 (0.6)	0.0 (0.1)	19.1 (8.5)	–
Index	–	–	10.2 (3.5)	8.9 (3.2)	37.6 (7.9)	18.8 (3.9)
Medial	–	–	10.2 (3.5)	0.0 (0.0)	37.6 (7.9)	18.8 (3.9)
Ring	0.0 (0.5)	–	6.2 (1.8)	9.8 (2.6)	22.8 (6.6)	11.4 (3.3)
Little	0.0 (1.0)	–	2.1 (0.7)	17.7 (4.5)	9.4 (3.8)	4.7 (1.9)
Cylinder 2						
Thumb	−10.0(3.1)	50.0 (4.6)	23.9 (0.3)	0.0 (0.1)	17.6 (3.3)	–
Index	–	–	2.4 (2.5)	10.0 (1.6)	38.1 (8.0)	19.0 (4.0)
Medial	–	–	2.4 (2.5)	0.0 (0.0)	38.1 (8.0)	19.0 (3.9)
Ring	0.0 (1.2)	–	1.6 (1.6)	11.6 (0.8)	24.2 (3.9)	12.1 (2.0)
Little	0.0 (0.7)	–	0.9 (0.8)	19.5 (2.1)	11.7 (2.2)	5.8 (2.9)

Table 5.5 Mean (SD) joint angles defining the grasping postures for both cylinders 1 and 2.

	MCC		MCP		PIP	DIP
	Flexion (°)	Abduction (°)	Flexion (°)	Abduction (°)	Flexion (°)	Flexion (°)
Cylinder 1						
Thumb	1.0 (0.5)	47.5 (2.7)	20.9 (0.2)	0.0 (0.3)	49.7 (6.9)	–
Index	–	–	26.2 (4.6)	−2.1(2.4)	48.3 (5.4)	24.1 (2.7)
Medial	–	–	26.2 (4.6)	0.0 (0.0)	48.3 (5.4)	24.1 (2.6)
Ring	4.0 (1.2)	–	19.5 (2.1)	7.1 (4.9)	35.6 (4.4)	17.8 (2.2)
Little	8.0 (0.9)	–	12.8 (1.4)	13.6 (8.4)	31.4 (5.7)	15.7 (2.9)
Cylinder 2						
Thumb	0.0 (0.7)	50.8 (2.6)	20.9 (0.3)	0.0 (0.8)	35.5 (8.4)	–
Index	–	–	9.4 (1.7)	9.7 (3.2)	46.7 (6.1)	23.4 (3.1)
Medial	–	–	9.4 (1.7)	0.0 (0.0)	46.7 (6.1)	23.4 (3.0)
Ring	4.0 (1.3)	–	7.6 (1.6)	9.1 (0.7)	33.6 (6.0)	16.8 (3.0)
Little	8.0 (0.8)	–	5.8 (1.7)	19.2 (1.4)	28.0 (5.6)	14.0 (2.8)

respectively. They are similar to the those measured, but not identical. Figure 5.31 shows the realistic appearance of the estimated grasping posture for cylinder 1.

The finger force patterns registered during the experiment can be observed in Fig. 5.32. This figure shows the forces registered by the Finger TPS system in one of the repetitions for cylinders 1 and 2. Due to the greater weight of cylinder 1, a peak is observed in the finger forces during the cylinder elevation phase corresponding to inertial effects, which is not observed for the case of cylinder 2. Finger forces registered during the static hold of the cylinders remain quite constant. For each repetition, the mean of the finger force registered during the static hold was considered. The mean of the finger forces among repetitions for both cylinders is shown in Table 5.7, along with the contact forces estimated by the model and three different modifications

Table 5.6 Grasping postures estimated by the model for both cylinders 1 and 2

	MCC		MCP		PIP	DIP
	Flexion (°)	Abduction (°)	Flexion (°)	Abduction (°)	Flexion (°)	Flexion (°)
Cylinder 1						
Thumb	0.3	47.7	21.2	0.0	47.3	–
Index	–	–	20.4	1.9	44.4	18.8
Medial	–	–	23.0	0.0	46.1	27.4
Ring	4.2	–	20.0	7.0	36.2	43.9
Little	8.5	–	13.5	13.4	32.8	7.8
Cylinder 2						
Thumb	−3.0	50.5	21.8	0.0	30.2	–
Index	–	–	9.7	9.7	47.1	23.5
Medial	–	–	12.9	0.0	51.1	25.5
Ring	6.7	–	11.6	7.4	40.0	20.0
Little	7.8	–	5.7	19.2	27.6	13.8

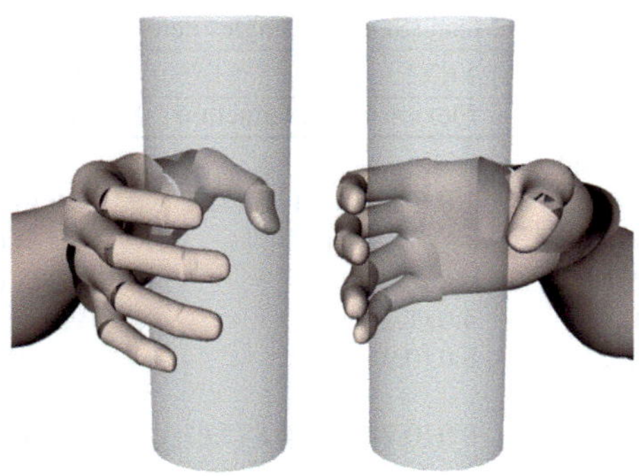

Fig. 5.31 Grasping posture estimated by the model for cylinder 1

are described later. It has to be noted that the forces registered for the thumb and the index finger varied greatly among repetitions. This confirms that there is not a unique combination of forces for grasping the cylinder, but that different safety margins can be applied for avoiding slipperiness of the object. Experimental mean contributions of the thumb and fingers to the grasp force are presented in Table 5.8, along with those estimated obtained from the model and its modifications. These contributions were calculated with respect to the sum of the contact grasping forces. The experimental contributions allow identifying the thumb as the major contributor to the grasping force.

Tables 5.7 and 5.8 compare the experimental finger forces registered during the static hold of the cylinders to the contact forces and torques estimated by the model and three different modifications. The original model (Estimated 1) considered the minimisation of the sum of the squared muscle stresses (maximisation of endurance) to solve the indeterminate problem, and a friction coefficient of 0.8, obtained from [75]. A disagreement between the experimentally measured normal forces and the

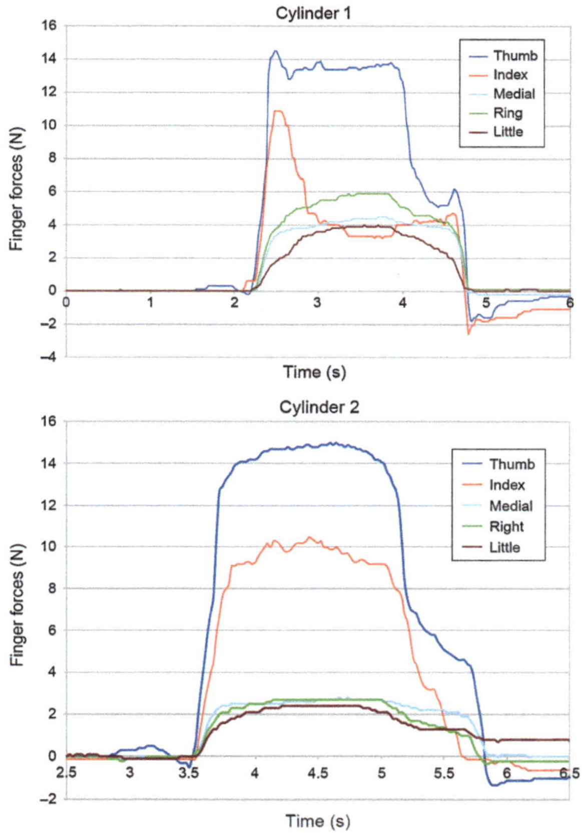

Fig. 5.32 Finger forces (N) registered during one of the repetitions for both cylinders 1 and 2

estimated forces is evident. On the one hand, the estimated values are lower than the experimental ones, being the total grasping force underestimated by 73 % in the case of holding cylinder 1, and by 97 % in the case of holding cylinder 2 (the lightest one). On the other hand, the estimated grasping force distributions among fingers do not match the force distributions measured experimentally. In particular, the model predicts that index and little fingers do not contribute at all to the grasp, which does not match the real behaviour of the human hand.

First modification to the model (Estimated 2) tries to investigate whether an improper friction coefficient between the hand and the cylinder could be responsible for the low level of the forces estimated by the model, given that a smaller friction coefficient will demand greater normal forces to assure the grasp stability. Taking into account the data reported by [75], the effect of changing this coefficient to a very low value of 0.3 was checked (Tables 5.7 and 5.8). Although the model estimates greater normal forces, the total grasping force estimated by the model for

Table 5.7 Mean (SD) finger forces registered for the grasping of both cylinders and contact forces estimated by the model and its modifications

	Thumb	Index	Medial	Ring	Little	Total
Cylinder 1						
Experimental						
Normal (N)	10.7 (4.5)	3.4 (2.4)	4.6 (0.8)	5.8 (0.6)	4.1 (1.1)	28.7 (3.2)
Estimated 1						
Normal (N)	4.05	0.01	1.97	0.96	0.60	7.59
Tangential (N)	3.23	0.01	1.57	0.76	0.48	
Torque (N· mm)	−4.70	0.00	−1.40	−0.60	−0.30	
Estimated 2						
Normal (N)	10.59	2.96	1.98	2.54	3.12	21.19
Tangential (N)	3.18	0.89	0.59	0.76	0.94	
Torque (N· mm)	0.00	0.10	−0.30	−0.50	−0.70	
Estimated 3						
Normal (N)	3.81	0.91	0.92	0.92	0.92	7.48
Tangential (N)	3.02	0.73	0.72	0.71	0.73	
Torque (N· mm)	−5.80	0.10	−1.10	−1.70	0.00	
Estimated 4						
Normal (N)	6.62	2.12	1.64	1.47	2.12	13.97
Tangential (N)	3.15	1.68	1.27	0.95	1.69	
Torque (N· mm)	−25.20	3.00	−3.80	−7.60	1.20	
Cylinder 2						
Experimental						
Normal (N)	11.0 (5.8)	9.3 (4.5)	3.0 (0.8)	2.9 (0.6)	1.7 (0.5)	27.9
Estimated 1						
Normal (N)	0.41	0.00	0.24	0.11	0.01	0.77
Tangential (N)	0.30	0.01	0.19	0.09	0.01	
Torque (N· mm)	1.00	0.00	−0.10	0.00	0.00	
Estimated 2						
Normal (N)	0.98	0.18	0.53	0.00	0.26	1.95
Tangential (N)	0.29	0.05	0.16	0.00	0.08	
Torque (N· mm)	−0.10	0.00	0.00	0.00	0.00	
Estimated 3						
Normal (N)	0.45	0.12	0.11	0.11	0.11	0.9
Tangential (N)	0.35	0.09	0.09	0.07	0.09	
Torque (N· mm)	−0.60	0.00	0.10	0.30	0.00	
Estimated 4						
Normal (N)	4.98	1.79	1.32	0.48	1.56	10.13
Tangential (N)	0.36	0.20	0.30	0.38	0.70	
Torque (N· mm)	−36.50	16.20	−10.90	0.90	−8.70	

grasping the lighter cylinder is still far from the registered force. Furthermore, a disagreement in the force contributions of the fingers remains, especially for the lighter cylinder. This seems to indicate that the underestimated results given by the original

Table 5.8 Mean finger force contribution (%) registered for the grasping of both cylinders and estimated by the model and its modifications

	Thumb (%)	Index (%)	Medial (%)	Ring (%)	Little (%)
Cylinder 1					
Experimental	37	12	16	20	14
Estimated 1	53	0	26	13	8
Estimated 2	50	14	9	12	15
Estimated 3	51	12	12	12	12
Estimated 4	47	15	12	11	15
Cylinder 2					
Experimental	39	33	11	10	6
Estimated 1	53	0	31	14	1
Estimated 2	50	9	27	0	13
Estimated 3	50	13	12	12	12
Estimated 4	49	18	13	5	15

model are not due to inaccuracies in the friction coefficient between the object and the hand.

The results from the simulations carried out with the original model (Estimated 1) seem to indicate that it is mathematically feasible to grasp the cylinders without the contribution of some fingers. And even that this fact could be more efficient in some aspects (maximising the endurance). However, the experimental results indicate that the CNS chooses a more even distribution of the forces between fingers. Trying to account for this coordination mechanism, we have proposed the second modification of the model (Estimated 3). We have repeated the simulations by adding to the objective function to be minimised a term accounting for the differences between the finger forces:

$$OBJ = \sum_{i=1}^{l} \left(\frac{F_i}{PCSA_i} \right)^2 + 100^2 \sum_{r,s=1}^{4} \left(F_{n_r} - F_{n_s} \right)^2 \Big|_{r \neq s} \quad (5.17)$$

where F_{n_r} is the normal component of the contact force developed by finger r and $PCSA$ is measured in centimetres. The factor 100^2 was introduced to make both combined functions comparable: we want to make the muscles stresses (approximately in the range of $10\,\mathrm{N/cm^2}$) to differences of forces of 0.1 N. The results from these comparable simulations are also presented in Tables 5.7 and 5.8. The use of this function allows achieving more balanced estimations of forces, but the magnitudes of the estimated forces are still lower than those of the experimental forces. The use of this objective function and the reduction in the friction coefficient to 0.3 (results not shown for brevity) provided a quite close estimation of forces for the heaviest cylinder, but the magnitudes of the estimated forces for the lightest cylinder were still too small compared with those of the experimental results.

All these results seem to point out that the criterion that the CNS uses to select the grasping force distribution among fingers is not only related to some energetic minimisation, as the experimental forces registered are much larger than those theoretically required to perform the grasp. The key must lie in some other factor. It is well known that humans exert grip forces taking a force safety margin into consideration to improve the grasp stability [98]. In this sense, in robotics, the selection of the grasp to be executed by a robotic hand is done by calculating different kinds of grasp quality measures. Many different quality measure definitions can be found in the robotics literature. Most of them are related to the capability of handling the object once grasped or the ability of the grasp to resist external disturbances (stability). This knowledge might be used also for studying the human grasp. For the experiment simulated in this study, it might make sense that the CNS would try to ensure certain level of stability of the cylinder being grasped, given that the subject was asked to hold it still for few seconds. The third modification of the original model (Estimated 4) was carried out to simulate this criterion.

Most of the robotic quality measures that evaluate the stability of the grasps are geometrical measures that only take into account the contact points and the directions of the normal contact forces. These measures do not account for the magnitudes of the forces and would not be useful for defining the objective function in the case under study. Obviously, the sum of the components of the applied forces that are normal to the object boundary is indicative of the force efficiency in the grasp. Then, a quality measure can be defined as the inverse of the sum of the magnitudes of the normal components of the applied forces required to balance an expected demanding wrench [99]. The index must be minimised to get an optimum grasp.

The results of minimising that function (Estimated 4), that looks for a more stable grasp, are also shown in Tables 5.7 and 5.8. The magnitudes of the forces estimated by the model with this assumption are much closer to the experimental forces than with any other of the previous objective functions, even for the lighter cylinder. These results confirm that, for the experiment being simulated, the CNS is trying to ensure the stability of the grasped cylinder. Although the results do not match exactly the experimental measurements, they adjust better than any other of the previously considered scenarios.

The differences between the estimated thumb and finger force values and the experimental ones can be also attributed to a deficiency in the kinematic model considered. [100] speculated that modelling the thumb CMC joint with two intersecting and orthogonal rotation axes could prevent to effectively replicate the contact forces produced by human subjects during maximum voluntary effort. This has been validated in a recent study by [101]. Therefore, a more exact kinematic model taking into account these studies should to be considered in future works to improve the model's estimations.

Anyway, the criterion selected by the CNS in each case should probably be a function of the task to be performed. The objective function that has provided good results in these simulations may not provide so good results under other requirements.

For example, if the subject were asked to grasp a cylindrical bottle to pour water. In that case, the grasp should allow certain level of manipulability that will be in conflict with the stability. More research is needed in this area. In any case, what seems clear is that the popular objective function sum of the squared muscle stresses is not suitable for grasping simulation using biomechanical models of the hand, or that it should be at least complemented by task-dependent grasp quality measures (manipulability or stability).

5.8 Conclusion

A realistic and self-contained biomechanical model of the hand has been developed by merging the current knowledge of biomechanics, ergonomics and robotics. The model simulates the complete hand and can be easily scaled to study different percentiles of populations. It has a realistic representation that allows the ergonomic evaluation of products. The model is dynamic and can be used to study the muscular patterns associated with a specific grasp. It allows predicting feasible grasping postures and provides the contact information required for evaluating the grasp. Finally, the model incorporates different quality grasping measures that are explained in the following chapters. All the abovementioned features are performed in a virtual environment developed using different software tools that have been chosen for their versatility, flexibility and open access.

There are several parts of the model that are open for improvement. Specifically, the skin is modeled as rigid bodies and a deformation algorithm for the finger joints could more realistically show the hand postures. Also, development of closure algorithms that do not need the most open and tentative grasp posture to be measured experimentally. Additionally, the user interface could also be improved to provide more options to the user.

The application shows how the simulation of the human hand biomechanics can be used to solve the indeterminate problem of finding the muscular forces adding the equilibrium of the grasped object, by minimising different objective functions. The model underestimated the normal contact forces when the criterion of minimising the sum of the squared muscle stresses was used. Furthermore, according to the model predictions, it is mathematically feasible to grasp the cylinders without the contribution of some fingers, and this is more efficient in some aspects. But it is not the real behaviour of the human hand that was experimentally observed.

For the simulated experiment, best results were obtained when the indeterminate problem was solved using a robotic grasp quality measure as objective function that tried to ensure the stability of the grasped cylinder. Although this function has provided good results in these simulations, it may fail for others entailing certain level of manipulability, as the criterion selected by the CNS in each case will probably be a function of the task to be performed. Further research on the application of other robotic grasp quality measures for different tasks involving different levels of stability and manipulability is needed.

Therefore, the main contribution of this application is in showing that the widespread Crowninshield minimisation function does not work well when trying to simulate the grasping of an object with an already validated 3D model of the hand. And that the consideration of stability criteria improves the estimations. This result is significant in the context of human grasp modelling and has not been reported previously in the literature, contributing to a better understanding of the human grasp.

Finally, the model presented in this study has been used to study only grasps of cylinders with the fingertips. More complex grasps, involving more contact zones and more complex object geometries should be investigated in the future.

References

1. Endo, Y., Kanai, S., Kishinami, T., Miyata, N., Kouchi, M., Mochimaru, M.: A computer-aided ergonomic assessment and product design system using digital hands. In: Proceedings of the 1st international conference on Digital human modeling. pp. 833–842. Springer, Germany (2007)
2. Endo, Y., Kanai, S., Miyata, N., Kouchi, M., Mochimaru, M., Konno, J., Ogasawara, M., Shimokawa, M.: Optimization-based grasp posture generation method of digital hand for virtual ergonomics assessment. SAE Int. J. Passeng. Cars-Electron. Electr. Syst. **1**(1), 590–598 (2008)
3. Goussous, F.A.: Grasp planning for digital humans. Ph.D. thesis, Iowa University, US (2007)
4. Kawaguchi, K.: Database-driven grasp synthesis and ergonomic assessment for handheld product design. Lecture Notes in Computer Science **5620**, 642–652 (2009)
5. Leijnse, J., Bonte, J.E., Landsmeer, J., Kalker, J.J., Vandermeulen, J.C., Snijders, C.: Biomechanics of the finger with anatomical restrictions - the significance for the exercising hand of the musician. J. Biomech. **25**(11), 1253–1264 (1992)
6. Storace, A., Wolf, B.: Functional-analysis of the role of the finger tendons. Journal of Biomechanics **12**(8), 575–578 (1979)
7. Guirintano, D.J., Hollister, A.M.: Force analysis of the thumb for a five-link system. In: ASME Biomechanics. Symposium. **120**, 213–217 (1991)
8. Esteki, A., Mansour, J.M.: A dynamic model of the hand with application in functional neuromuscular stimulation. Ann. Biomed. Eng. **25**(3), 440–451 (1997)
9. Weightman, B., Amis, A.A.: Finger joint force predictions related to design of joint replacements. Journal of Biomedical Engineering **4**(3), 197–205 (1982)
10. Fok, K.S., Chou, S.M.: Development of a finger biomechanical model and its considerations. J. Biomech. **43**(4), 701–713 (2010)
11. Kamper, D., Fischer, H., Cruz, E.: Impact of finger posture on mapping from muscle activation to joint torque. Clin. Biomech. **21**(4), 361–369 (2006)
12. Kurita, Y., Onoue, T., Ikeda, A.: Ogasawara. T, Biomechanical analysis of subjective pinching effort based on tendon-skeletal model (2009)
13. Lee, J.W., Rim, K.: Maximum finger force prediction using a planar simulation of the middle finger. Proc. Instn. Mech. Engnrs. Part H. J. Eng. Med. **204**, 169–178 (1990)
14. Lee, S.W., Chen, H., Towles, J.D., Kamper, D.G.: Effect of finger posture on the tendon force distribution within the finger extensor mechanism. J. Biomech. Eng. Trans. Asme **130**(5), 051014 (2008)
15. Lee, S.W., Chen, H., Towles, J.D., Kamper, D.G.: Estimation of the effective static moment arms of the tendons in the index finger extensor mechanism. J. Biomech. **41**(7), 1567–1573 (2008)
16. Qiu, D., Fischer, H.C.: Kamper. muscle activation patterns during force generation of the index finger, D.G. (2009)

17. Roloff, I., Schoffl, V., Vigouroux, L., Quaine, F.: Biomechanical model for the determination of the forces acting on the finger pulley system. J. Biomech. **39**(5), 915–923 (2006)
18. Sancho-Bru, J.L., Perez-Gonzalez, A., Vergara-Monedero, M., Giurintano, D.: A 3-d dynamic model of human finger for studying free movements. J. Biomech. **34**(11), 1491–1500 (Nov 2001)
19. Sancho-Bru, J.L., Giurintano, D.J., Pérez-González, A., Vergara, M.: Optimum tool handle diameter for a cylinder grip. J. Hand Ther.: Official J. Am. Soc. Hand Ther. **16**(4), 337–342 (2003)
20. Sancho-Bru, J.L., Perez-Gonzalez, A., Vergara, M., Giurintano, D.J.: A 3d biomechanical model of the hand for power grip. J. Biomech. Eng. **125**(1), 78–83 (2003)
21. Sancho-Bru, J., Vergara, M., Rodríguez-Cervantes, P.J., Giurintano, D., Pérez-González, A.: Scalability of the muscular action in a parametric 3d model of the index finger. Ann. Biomed. Eng. **36**, 102–107 (2008)
22. Valero-Cuevas, F.J.: Predictive modulation of muscle coordination pattern magnitude scales fingertip force magnitude over the voluntary range. J. Neurophysiol. **83**(3), 1469–1479 (2000)
23. Valero-Cuevas, F.J.: An integrative approach to the biomechanical function and neuromuscular control of the fingers. J. Biomech. **38**(4), 673–684 (2005)
24. Vigouroux, L., Quaine, F., Labarre-Vila, A., Moutet, F.: Estimation of finger muscle tendon tensions and pulley forces during specific sport-climbing grip techniques. J. Biomech. **39**(14), 2583–2592 (2006)
25. Vigouroux, L., Ferry, M., Colloud, F., Paclet, F., Cahouet, V., Quaine, F.: Is the principle of minimization of secondary moments validated during various fingertip force production conditions? Hum. Mov. Sci. **27**(3), 396–407 (2008)
26. Wu, J.Z., An, K.N., Cutlip, R.G., Dong, R.G.: A practical biomechanical model of the index finger simulating the kinematics of the muscle/tendon excursions. Biomed. Mater. Eng. **20**(2), 89–97 (2010)
27. Brand, P., Hollister, A.: Clinical mechanics of the hand. Elsevier Science Health Science div (1992)
28. Giurintano, D.J., Hollister, A.M., Buford, W.L., Thompson, D.E., Myers, L.M.: A virtual five-link model of the thumb. Med. Eng. Phys. **17**(4), 297–303 (1995)
29. Youm, Y., Gillespie, T., Flatt, A., Sprague, B.: Kinematic investigation of normal mcp joint. J. Biomech. **11**(3), 109–118 (1978). http://www.sciencedirect.com/science/article/pii/0021929078900039
30. An, K.N., Chao, E.Y., Cooney, W.P., Lincheid, R.L.: Normative model of human hand for biomechanical analysis. J. Biomech. **12**(10), 775–788 (1979)
31. Crowninshield, R.D., Brand, R.A.: A physiologically based criterion of muscle force prediction in locomotion. J. Biomech. **14**(11), 793–801 (1981)
32. Colombo, G., Cugini, U.: Virtual humans and prototypes to evaluate ergonomics and safety. J. Eng. Des. **16**(2), 195–203 (2005). http://www.tandfonline.com/doi/abs/10.1080/09544820500031542
33. Yang, J., Kim, J.H., Abdel-Malek, K., Marler, T., Beck, S., Kopp, G.R.: A new digital human environment and assessment of vehicle interior design. Comput. Aided Des. **39**(7), 548–558 (2007). http://www.sciencedirect.com/science/article/pii/S0010448506002120, human Modeling and Applications
34. Case, K., Porter, J., Bonney, M.: SAMMIE: a man and workplace modeling system. Comput. Aided Ergono., 31–56 (1990)
35. Peña Pitarch, E.: Virtual human hand: Grasping strategy and simulation. Ph.D. thesis, Universitat Politècnica de Catalunya (2007)
36. Mital, A., Kilbom, A.: Design, selection and use of hand tools to alleviate trauma of the upper extremities: Part ii-the scientific basis (knowledge base) for the guide. In: Anil Mital, A.K., Kumar, S. (eds.) Ergonomics Guidelines and Problem Solving, Elsevier Ergonomics Book Series, vol. 1, pp. 217–231. Elsevier, Amsterdam (2000). http://www.sciencedirect.com/science/article/pii/S1572347X00800178

37. Keyserling, W.M.: Workplace risk factors and occupational musculoskeletal disorders, part 2: A review of biomechanical and psychophysical research on risk factors associated with upper extremity disorders. AIHAJ **61**(2), 231–243 (2000). http://www.ncbi.nlm.nih.gov/pubmed/10782195
38. Kong, Y., Jang, H., Freivalds, A.: Wrist and tendon dynamics as contributory risk factors in work-related musculoskeletal disorders: Research articles. Hum. Factor. Ergon. Manuf. **16**(1), 83–105 (2006). http://dx.doi.org/10.1002/hfm.v16:1
39. Muggleton, J.M., Allen, R., Chappell, P.H.: Hand and arm injuries associated with repetitive manual work in industry: a review of disorders, risk factors and preventive measures. Ergonomics **42**(5), 714–739 (1999). http://www.tandfonline.com/doi/abs/10.1080/001401399185405
40. Schoenmarklin, R.W., Marras, W.S., Leurgans, S.E.: Industrial wrist motions and incidence of hand/wrist cumulative trauma disorders. Ergonomics **37**(9), 1449–1459 (1994). http://www.ncbi.nlm.nih.gov/pubmed/7957023
41. Spielholz, P., Silverstein, B., Morgan, M., Checkoway, H., Kaufman, J.: Comparison of self-report, video observation and direct measurement methods for upper extremity musculoskeletal disorder physical risk factors. Ergonomics **44**(6), 588–613 (2001). http://www.ncbi.nlm.nih.gov/pubmed/11373023
42. Kumar, S.: Theories of musculoskeletal injury causation. Ergonomics **44**(1), 17–47 (2001). http://www.ncbi.nlm.nih.gov/pubmed/11214897
43. Armstrong, T.J., Best, C., Bae, S., Choi, J., Grieshaber, D.C., Park, D., Woolley, C., Zhou, W.: Development of a kinematic hand model for study and design of hose installation. Lect. Notes Comput. Sci. **5620**, 85–94 (2009)
44. Freivalds, A.: Biomechanics of the upper limbs: Mechanics, modeling, and musculoskeletal Injuries (2004)
45. Wu, G., van der Helm, F.C., Veeger, H.D., Makhsous, M., Roy, P.V., Anglin, C., Nagels, J., Karduna, A.R., McQuade, K., Wang, X., Werner, F.W., Buchholz, B.: Isb recommendation on definitions of joint coordinate systems of various joints for the reporting of human joint motion-part ii: shoulder, elbow, wrist and hand. J. Biomech. **38**(5), 981–992 (2005). http://www.sciencedirect.com/science/article/pii/S002192900400301X
46. Kapandji, A.I.: Fisiologie articulaire. Membre Supérieur. Editions Maloine, Paris (1998)
47. Buchholz, B., Armstrong, T.J.: A kinematic model of the human hand to evaluate its prehensile capabilities. J. Biomech. **25**(2), 149–162 (1992)
48. Hollister, A., Giurintano, D.J., Buford, W.L., Myers, L.M., Novick, A.: The axes of rotation of the thumb interphalangeal and metacarpophalangeal joints. Clin. Orthop. Relat. Res. **320**, 188–193 (1995)
49. Sancho-Bru, J.: Model biomecànic de la mà orientat al disseny d'eines manuals. Ph. d. thesis, Universitat Jaume I (2000)
50. Tubiana, R.: The hand, vol. 1. Sanders Company, Philadelphia (1981)
51. Tubiana, R., T.J., Mackin, E.: Examination of the hand and wrist. Martin Dunitz (1996)
52. Denavit, J.: A kinematic notation for lower-pair mechanisms based on matrices. Trans. ASME. J. Appl. Mech. **22**, 215–221 (1955). http://ci.nii.ac.jp/naid/10008019314/en/
53. Lee, K.S., Mo, S.M., Hwang, J.J., Wang, H., Jung, M.C.: Relaxed hand postures. Japanese J. Ergon. **44**(Supplement), 436–439 (2008)
54. An, K.N., Chao, E.Y.S., Kaufman, K.R.: Analysis of muscle and joint loads, pp. 1–50. Basic Orthopaedics Biomechanics, Raven Press. Ltd., New York (1991).
55. Goldstein, S.A., Armstrong, T.J., Chaffin, D.B., Matthews, L.S.: Analysis of cumulative strain in tendons and tendon sheaths. J. Biomech. **20**(1), 1–6 (1987)
56. Kaufman, K.R., An, K.N., Litchy, W.J., Chao, E.Y.S.: Physiological prediction of muscle forces.1. theoretical formulation. Neuroscience **40**(3), 781–792 (1991)
57. Hatze, H.: Myocibernetic control models of skeletal muscle (1981)
58. Valero-Cuevas, F.J., Zajac, F.E., Burgar, C.G.: Large index-fingertip forces are produced by subject-independent patterns of muscle excitation. J. Biomech. **31**(8), 693–703 (1998)

59. Zajac, F.E.: Muscle and tendon - properties, models, scaling, and application to biomechanics and motor control. Crit. Rev. Biomed. Eng. **17**(4), 359–411 (1989)
60. Lemay, M.A., Crago, P.E.: A dynamic model for simulating movements of the elbow, forearm, and wrist. J. Biomech **29**(10), 1319–1330 (1996)
61. Jacobson, M.D., Raab, R., Fazeli, B.M., Abrams, R.A., Botte, M.J., Lieber, R.L.: Architectural design of the human intrinsic hand muscles. J. Hand Surg. Am. **17A**(5), 804–809 (1992)
62. Eyler, D.L., Markee, J.E.: The anatomy and function of the intrinsic musculature of the fingers. J. Bone Joint Surg. **36**(1), 1–18 (1954)
63. Dubousset, J.: The digital joints, pp. 191–201. Saunders Company, Philadelphia (1981)
64. Craig, S.M.: Anatomy of the joints of the fingers. Hand Clin. **8**(4), 693–700 (1992). http://www.ncbi.nlm.nih.gov/pubmed/1460067
65. Mommersteeg, T., Blankevoort, L., Huiskes, R., Kooloos, J., Kauer, J.: Characterization of the mechanical behavior of human knee ligaments: A numerical-experimental approach. J. Biomech. **29**(2), 151–160 (1996). http://www.sciencedirect.com/science/article/pii/0021929095000402
66. Minami, A., An, K.N., Cooney, W.P., Linscheid, R.L., Chao, E.Y.: Ligament stability of the metacarpophalangeal joint: a biomechanical study. J. Hand Surg. Am. **10**(2), 255–260 (1985). http://www.ncbi.nlm.nih.gov/pubmed/3980940
67. Ciocarlie, M., Miller, A., Allen, P.: Grasp analysis using deformable fingers. IEEE/RSJ International Conference on Intelligent Robots and Systems, In (2005). (IROS 2005). pp. 4122–4128 (2005)
68. Ciocarlie, M., Lackner, C., Allen, P.: Soft finger model with adaptive contact geometry for grasping and manipulation tasks. Second Joint EuroHaptics Conference and Symposium on Haptic Interfaces for Virtual Environment and Teleoperator Systems, pp. 219–224 (2007)
69. Howe, R., Kao, I., Cutkosky, M.: The sliding of robot fingers under combined torsion and shear loading. In: Proceedings of IEEE International Conference on Robotics and Automation, vol. 1. pp. 103–105 (1988)
70. Kao, I., Cutkosky, M.R.: Quasistatic manipulation with compliance and sliding. Int. J. Robot. Res. **11**(1), 20–40 (1992)
71. Howe, R.D., Cutkosky, M.R.: Practical force-motion models for sliding manipulation. I. J. Robot. Res. **15**(6), 557–572 (1996). http://dblp.uni-trier.de/db/journals/ijrr/ijrr15.html
72. Kao, I., Yang, F.: Stiffness and contact mechanics for soft fingers in grasping and manipulation. IEEE Trans. Robot. Autom. **20**(1), 132–135 (2004)
73. Johnson, K.L.: Contact mechanics. Cambridge University Press, Cambridge (1995)
74. Derler, S., Gerhardt, L.C.: Tribology of skin: Review and analysis of experimental results for the friction coefficient of human skin. Tribol. Lett. **45**, 1–27 (2012). http://dx.doi.org/10.1007/s11249-011-9854-y
75. Savescu, A.V., Latash, M.L., Zatsiorsky, V.M.: A technique to detershrane friction at the fingertips. J. Appl. Biomech. **24**(1), 43–50 (2008)
76. Han, H.Y., Kawamura, S.: Analysis of stiffness of human fingertip and comparison with artificial fingers. In: Systems, Man, and Cybernetics, 1999. IEEE SMC '99 Conference Proceedings. 1999 IEEE International Conference on. vol. 2, pp. 800–805 (1999)
77. Davidoff, N.A., Freivalds, A.: A graphic model of the human hand using CATIA. Int. J. Ind. Ergon. **12**(4), 255–264 (1993). http://dx.doi.org/10.1016/0169-8141(93)90095-U
78. Bernabeu, E., Tornero, J.: Hough transform for distance computation and collision avoidance. IEEE Trans. Robot. Autom. **18**(3), 393–398 (2002)
79. Gilbert, E.G., Johnson, D.W., Keerthi, S.S.: A fast procedure for computing the distance between complex objects in 3-dimensional space. IEEE J. Robot. Autom. **4**(2), 193–203 (1988)
80. Rhee, T., Neumann, U., Lewis, J.P.: Human hand modeling from surface anatomy. In: Proceedings of the 2006 symposium on Interactive 3D graphics and games. pp. 27–34. I3D '06, ACM, New York, NY, USA (2006). http://doi.acm.org/10.1145/1111411.1111417
81. Rogers, M.S., Barr, A.B., Kasemsontitum, B., Rempel, D.M.: A three-dimensional anthropometric solid model of the hand based on landmark measurements. Ergonomics **51**(4), 511–526 (2008). http://www.tandfonline.com/doi/abs/10.1080/00140130701710994

82. van Nierop, O.A., van der Helm, A., Overbeeke, K.J., Djajadiningrat, T.J.P.: A natural human hand model. Vis. Comput. **24**(1), 31–44 (2007). http://dx.doi.org/10.1007/s00371-007-0176-x
83. Mora, M., Sancho-Bru, J.L., Pérez-González, A.: Hand posture prediction using neural networks within a biomechanical model. Int. J. Adv. Robot, Syst (2012)
84. Choi, J.: Developing a three-dimensional kinematic model of the hand for ergonomic analyses of hand posture, hand space envelope, and tendon excursion. Ph.D. thesis, University of Michigan (2008).
85. de Jalón, J.G., Bayo, E.: Kinematic and dynamic simulation of multibody systems. Springer, New York (1994)
86. Tsang, W., Singh, K., Fiume, E.: Helping hand: an anatomically accurate inverse dynamics solution for unconstrained hand motion. In: Proceedings of the 2005 ACM SIGGRAPH/Eurographics symposium on Computer animation. pp. 319–328. SCA '05, ACM, USA (2005). http://doi.acm.org/10.1145/1073368.1073414
87. Albrecht, I., Haber, J., Seidel, H.P.: Construction and animation of anatomically based human hand models. In: Proceedings of the 2003 ACM SIGGRAPH/Eurographics symposium on Computer animation. pp. 98–109. SCA '03, Eurographics Association, Aire-la-Ville, Switzerland (2003). http://dl.acm.org/citation.cfm?id=846276.846290
88. Delp, S.L., Loan, J.: A graphics-based software system to develop and analyze models of musculoskeletal structures. Comput. Biol. Med. **25**(1), 21–34 (1995). http://www.sciencedirect.com/science/article/pii/001048259598882E
89. Delp, S., Loan, J.: A computational framework for simulating and analyzing human and animal movement. Comput. Sci. Eng. **2**(5), 46–55 (2000). http://ieeexplore.ieee.org/xpl/articleDetails.jsp?reload=true&arnumber=877394&contentType=Journals+%26+Magazines
90. Gonzalez, R.V., Buchanan, T.S., Delp, S.L.: How muscle architecture and moment arms affect wrist flexion-extension moments. J. Biomech. **30**(7), 705–712 (1997). http://www.sciencedirect.com/science/article/pii/S0021929097000158
91. Van Der Helm, F.C.: A three-dimensional model of the shoulder and elbow. In: Proceedings of the First Conference of the International Shoulder Group. pp. 65–70 (1997).
92. Holzbaur, K.R.S., Murray, W.M., Delp, S.L.: A model of the upper extremity for simulating musculoskeletal surgery and analyzing neuromuscular control. Ann. Biomed. Eng. **33**(6), 829–840 (2005). http://dx.doi.org/10.1007/s10439-005-3320-7
93. Delp, S.L., Anderson, F.C., Arnold, A.S., Loan, P., Habib, A., John, C.T., Guendelman, E., Thelen, D.G.: Opensim: Open-source software to create and analyze dynamic simulations of movement. IEEE Trans. Biomed. Eng. **54**(11), 1940–1950 (2007). http://dblp.uni-trier.de/db/journals/tbe/tbe54.html
94. Miller, A., Allen, P.: Graspit! a versatile simulator for robotic grasping. IEEE Robot. Autom. Mag. **11**(4), 110–122 (2004)
95. Diankov, R.: Automated Construction of Robotic Manipulation Programs. Ph.D. thesis, Carnegie Mellon University, Robotics Institute (2010)
96. Mora, M., Sancho-Bru, J.L., Iserte, J.L., Pérez-González, A.: Protocolo de calibración para guante instrumentado en la caracterización cinemática del agarre humano. X Congreso Iberoamericano de Ingeniería Mecánica, In (2011)
97. Sancho-Bru, J., Vergara, M., J.B.N.J., Mora Aguilar, M., Pérez-González, A.: Medición del movimiento de todos los segmentos de la mano mediante videogrametría. In: XIX Congreso Nacional de Ingeniería Mecánica (CNIM 2012). Castellon, Spain (2012).
98. Jenmalm, P., Goodwin, A.W., Johansson, R.S.: Control of grasp stability when humans lift objects with different surface curvatures. J. Neurophysiol. **79**(4), 1643–1652 (1998). http://www.ncbi.nlm.nih.gov/pubmed/9535935
99. Liu, G., Li, Z.: Real-time grasping-force optimization for multifingered manipulation: theory and experiments. IEEE/ASME Trans. Mechatron. **9**(1), 65–77 (2004)
100. Valero-Cuevas, F.J., Johanson, M.E., Towles, J.D.: Towards a realistic biomechanical model of the thumb: The choice of kinematic description may be more critical than the solution method

or the variability/uncertainty of musculoskeletal parameters. J. Biomech. **36**(7), 1019–1030 (2003)
101. Wohlman, S.J., Murray, W.M.: Bridging the gap between cadaveric and in vivo experiments: A biomechanical model evaluating thumb-tip endpoint forces. J. Biomech. **46**(5), 1014–1020 (2013). http://www.sciencedirect.com/science/article/pii/S0021929012006513

Chapter 6
Human Grasp Evaluation

6.1 Introduction

As it was presented in the previous chapter, many biomechanical human hand models have been developed so far, with the aim of providing a tool for studying problems that cannot be directly analysed on humans or that have a too high cost. One of the main features of the human hand is its grasping capability. However, the current models have a limited ability to predict feasible grasping postures and do not allow the evaluation of the quality of grasps.

Evaluating the quality of a human grasp could have several applications. First of all, inside biomechanical human hand models, it could be used as a tool for assisting in the prediction of grasping postures or as a criterion to solve the indeterminate problem of finding the contact forces to grasp an object in a given posture as shown in Sect. 5.7 [1]. Second, it can be applied in the design of hand-held products [2, 3]. Additionally, the design of hand prosthesis could also be improved if the quality of the grasp performed by a given mechanical hand could be measured and compared to the grasp performed by the physiological hand. Therefore, having a model that incorporates grasp quality measures might significantly increase their use by the biomechanics, medical and ergonomics communities.

For many years the robotics community has been studying the autonomous handling of objects by robots. In order to help the selection of the proper robotic grasp for handling an object, many grasp quality measures have been developed that allow comparing different aspects of the robotic grasp [4]. Although the human hand is obviously more complex than robot hands, the methods used in robotics might be adopted to study the human grasp. There are few studies evaluating the quality of a human grasp [2, 3]. Both works used the robotic measure of stability proposed by [5]. To the best of our knowledge, there has not been any study adapting other robotic quality measures to the human hand or proposing a global human grasp quality index.

B. León et al., *From Robot to Human Grasping Simulation*,
Cognitive Systems Monographs 19, DOI: 10.1007/978-3-319-01833-1_6,

6.2 Grasp Quality Measures: Literature Review

Many approaches have been proposed to measure the quality of a grasp. Some of the measures focus on evaluating the ability to resist external disturbances, others on evaluating the dexterity. In order to find the optimal grasp for a given task, the geometry of the object and the structure of the hand should be taken into account. Therefore, object, hand and task constraints should be considered when searching for the optimal grasp points (see Fig. 6.1).

Different measures have been proposed in robotics which takes one or more of these constraints into account. Object and hand constraints have been considered using the contact points to evaluate the quality of a grasp, using either the Grasp Matrix G, the location of the contact points or the contact forces that should be applied to resist external disturbances. Additionally, hand constraints have also been considered measuring the comfort of the grasp or the ability to manipulate the object in a given hand configuration. Finally, task constraints have also been added favoring the grasps that are able to generate wrenches on the object that are relevant to the task.

On the other hand, having the design of the object into account, some ergonomic quality measures have been proposed which evaluate the ease of grasping it or the level of risk factors associated with specific manipulation tasks.

A literature review of the proposed robotic and ergonomic grasp quality measures is presented in the following sections.

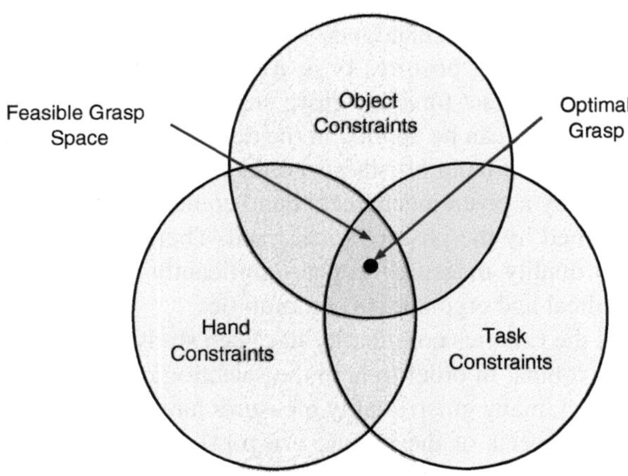

Fig. 6.1 Optimal grasp of multi-fingered hand adapted from [6]

6.2.1 Robotic Measures of Grasp Performance

The robotic quality measures can be classified into six groups depending on whether they consider the algebraic properties of G; the distribution of the contact points on the object; the magnitude of the forces applied on this contacts; the configuration of the manipulator; the task to be performed after the grasp; or a combination of them.

6.2.1.1 Algebraic Properties of G

These are stability indicators that consider the algebraic properties of the grasp matrix G to measure the grasp capability of withstanding external wrenches; they use the contact points and normal directions. They do not consider any limitation on the finger forces, so that in some cases the fingers have to apply very large forces to resist small perturbations.

The grasp matrix G (defined in Sect. 2.3.1) is the linear transformation between the contact force vector and the net wrench on the object. Geometrically, G maps the unitary sphere in the force domain of the contact points into an ellipsoid in the wrench space [7]. The lengths of the principal axes of the ellipsoid are the singular values of G, which are the n nonnegative square roots σ_i of the eigenvalues of GG^T. The following quality measures use these singular values to define the quality of the grasp.

Smallest singular value of G [8]

Objective: Maximize to obtain an optimum grasp

Description: This quality measure indicates how far the grasp configuration is from falling into a singular configuration, losing the capability of withstanding external wrenches [8]. When a grasp is in a singular configuration, at least one of the singular values of G is zero. The quality measure is defined as:

$$Q = \sigma_{min}(G) \tag{6.1}$$

where $\sigma_{min}(G)$ is the smallest singular value of the matrix G

Units: None

Lower limit: 0

Upper limit: Not determined

Advantages: Its computational simplicity

Limitations: It is invariant under body coordinate transformation but not invariant under a change of the toque origin. Also, since it gives only a worst case analysis it does not reflect the entire grasp

---

Volume of G in the wrenchspace [8]

Objective: Maximize to obtain an optimum grasp
Description: The global contribution of all the contact forces can be considered calculating the volume of the ellipsoid in the wrench space [7]
The volume of a matrix can be obtained as the product of its singular values [9], therefore this quality measure can be defined as:

$$Q = v(G) = \prod_{i=1}^{r} \sigma_i \tag{6.2}$$

where r is the rank of G, and $\sigma_1 \geq \sigma_2 \geq \cdots \geq \sigma_r$ denote the nonzero singular values of G
Units: None
Lower limit: 0
Upper limit: Not determined
Advantages: It is invariant under both a body coordinate transformation and a change of the origin
Limitations: It does not reflect stability of a grasp. There could be a case where an unstable grasp might yield a larger volume measure than a stable grasp [8]. Consequently, to compare the qualities of two grasps, first their stability should be checked and then use this measure to evaluate their quality

---

Grasp Isotropy Index [6]

Objective: Maximize to obtain an optimum grasp
Description: The precise position or force control may not be guaranteed in part if any finger lies near a singular position. This quality measure tries to obtain an isotropic grasp where the magnitude of internal forces are similar [6], and is calculated as:

$$Q = \frac{\sigma_{min}(G)}{\sigma_{max}(G)} \tag{6.3}$$

where σ_{min} and σ_{max} denote the minimum and maximum singular values of G. This measure approaches to one at a desirable configuration (isotropic) and is equal to zero at the singular configuration
Units: None
Lower limit: 0
Upper limit: 1
Advantages: It is normalized between 0 and 1. Also, its computational simplicity
Limitations: None

6.2.1.2 Distribution of the Contact Points

These are stability indicators that use the location of the contact points. Better stability of the grasps is assumed when contact points are distributed in a uniform way on the object surface. This can be calculated by measuring either the angles or the area of the polygon whose vertices are the contact points. Another measure considers the distance between the centroid of the contact points polygon and the object center of mass aiming to minimize the effect of gravitational and inertial forces.

Distance between the centroid of the contact polygon and the object's center of mass [10, 11]

Objective: Minimize to obtain an optimum grasp

Description: This index aims to minimize the effect of gravitational and inertial forces during the motion of the robot, measuring the distance between the center of mass p of the grasped object and the centroid of the contact points p_c [10]. The centroid of the contact points is calculated as:

$$p_c = \frac{1}{n_c} \sum_{i=1}^{n_c} c_i \tag{6.4}$$

where n_c is the number of contact points and c_i is the location of each contact point. Then the measure is calculated as:

$$Q = distance(p, p_c) \tag{6.5}$$

where p can be calculated as the centroid of the object when it can be assumed that the object has a uniform mass distribution. The smaller this distance the better the grasp can resist the effect of external forces

Units: $[length]$

Lower limit: 0

Upper limit: Maximum distance between the object's centroid and any point on the object surface

Advantages: It can be used for more than one contact, including more than three non-coplanar contacts

Limitations: None

Area of the grasp polygon[12]

Objective: Maximize to obtain an optimum grasp

Description: This measure is defined as the area of the polygon formed by the contact points. The optimum grasp is the one that best resists forces and torques about the grip plane, which has the largest polygon area [13]. It has been used in robotics for three finger hands, where all finger points lie in a plane [14]

$$Q = Area(Polygon(c_1, \ldots, c_i, \ldots, c_n)) \tag{6.6}$$

Units: $[length]^2$
Lower limit: 0
Upper limit: Not determined
Advantages: The area of the polygon should be more important to measure for heavy objects, where stability against gravitational and inertial torques is a critical issue [15]
Limitations: Only for contact points that lay one the same plane, therefore in robotics it has only been tested for two and three-finger planar grasps

---

Shape of the grasp polygon [6]

Objective: Minimize to obtain an optimum grasp
Description: This measure compares how far the internal angles of the grasp polygon are from those of the corresponding regular polygon [6]. According to [16], the quality of a grasp is improved in terms of stability, sliding avoidance and force equilibrium, when the contact points are distributed in a uniform way on the object surface, generating a regular polygon. This index is calculated as:

$$Q = \frac{1}{\theta_{max}} \sum_{i=1}^{n_f} |\theta_i - \bar{\theta}| \tag{6.7}$$

where n_f denotes the number of fingers, θ_i is the inner angle at the ith vertex of the polygon, $\bar{\theta}$ denotes the average angle of all inner angles of the grasp polygon, given by:

$$\bar{\theta} = \frac{180(n_f - 2)}{n_f} \tag{6.8}$$

and θ_{max} is the sum of the differences between the internal angles when the polygon has the most ill-conditioned shape (such as a line) and those of a regular polygon:

$$\theta_{max} = \sum_{i=1}^{n_f} |\theta_i - \bar{\theta}|_{ill\ conditioned} = (n_f - 2)(180 - \bar{\theta}) + 2\bar{\theta} \tag{6.9}$$

Note that in such a condition, the polygon would have two of the angles close to zero and the others close to 180°. The division of Q by θ_{max} aims to normalize the measure between zero and one
Units: None
Lower limit: 0

Upper limit: 1
Advantages: Its computational simplicity
Limitations: As the previous measure, defined only for contact points that lay on the same plane. Also, there could be many polygons with the same quality, therefore another grasp index is necessary to classify more stable grasp points.

Minimum length of Independent Contact Regions (ICR) [17]

Objective: Maximize to obtain an optimum grasp
Description: This measures take into account the uncertainty in the position of the fingers minimizing the sensitivity of a grasp to positioning errors because of the uncertainty in the robotic systems. Therefore, instead of contact points it calculates independent contact regions (*ICR*) in which force closure grasps are assured. The quality of the grasp is measured by the size of these regions, aiming to maximize the minimum of their lengths

$$Q = min_{r \in ICR} L_r \tag{6.10}$$

where L_r is the length in 2D space of the region
Units: [*length*]
Lower limit: 0
Upper limit: Not determined
Advantages: Considers the sensitivity of a grasp to positioning errors
Limitations: It has been applied only to 2D objects and two and three-finger grasps

---

Uncertainty grasp index[6]

Objective: Minimize to obtain an optimum grasp
Description: This index aims to estimate the position sensitivity of the grasp points, which is minimized if all fingers grasp the object at the center of the independent contact regions. It is calculated as the sum of the distances between each contact point $c_i = (x_i, y_i, z_i)$ and the center of the corresponding independent contact region (x_{io}, y_{io}, z_{io})

$$Q = \frac{1}{n_c} \sum_{i=1}^{n_c} \sqrt{(x_i - x_{io})^2 + (y_i - y_{io})^2 + (z_i - z_{io})^2} \tag{6.11}$$

Units: [*length*]
Lower limit: 0
Upper limit: Not determined
Advantages: Considers the sensitivity of a grasp to positioning errors
Limitations: Applicable only to planar grasps. The independent contact regions need to be calculated

---

6.2.1.3 Magnitude of Forces

These are stability indicators that take into account the magnitude of forces applied at the contact points. The previous indicators do not consider any limitation on the finger forces, so that in some cases the fingers have to apply very large forces to resist small perturbations.

Smallest maximum wrench to be resisted [5, 18]

Objective: Maximize to obtain an optimum grasp

Description: This measure defines the quality of a grasp as the length of the smallest wrench (in any direction) that breaks the grasp, when in every contact a force with unit strength is applied [19]

Better grasp configurations are assumed if any external force can be balanced with small finger forces, with small deformation of the object and fingers. Therefore, the quality of a grasp can be measured considering the ratio between the magnitude of the maximum wrench to be resisted (over all possible directions) and the magnitude of the applied finger forces. The magnitude of a wrench is defined as:

$$\|w\| = \sqrt{\|f\|^2 + \rho\|\tau\|^2} \tag{6.12}$$

where f and τ are the force and torque vectors and ρ is a parameter used to scale the torque to the force magnitude (see Sect. 2.2.1)

The *local quality measure* (*LQ*) is defined as:

$$LQ_w = \max_{g \in wA} \frac{\|w\|}{\|g\|} \tag{6.13}$$

where wA is the set of generalized forces that can resist the wrench w and g is the finger applied force for a given wrench direction. Without loss of generality, $\|w\|$ can be chosen so that $\|g\| = 1$ (we are considering the ratio and both scale linearly with each other). Therefore, we can rewrite LQ as:

$$LQ_{\hat{w}} = \max_{w \in wB} \|w\| \tag{6.14}$$

where wB is the set of generalized forces that can resist the wrench w such that $\|g\| = 1$

The *grasp quality measure* Q can be defined, for a set of contacts on the object, as:

$$Q = \min_{\hat{w}} LQ_{\hat{w}} = \min_{\hat{w}} \max_{w \in wB} \|w\| \tag{6.15}$$

The minimum is taken because there is usually no control over the wrench that the gripper must resist, then taken the minimum is a way to ensure a level of performance over all possible wrenches

The Grasp Wrench Space (GWS) is the set of all wrenches that can be applied to the object through the grasp contacts. A way to approximate the GWS is to calculate the convex hull over a discretized frictions cones. Let the convex hull of the primitive wrenches (CW) be the boundary of the set of wrenches that can be resisted by the applied forces. Then Q can be written as:

$$Q = \min_{w \in CW} \|w\| \tag{6.16}$$

This can be geometrically interpreted as the distance of the nearest point to the origin of the wrench-space, from the origin itself (the closest facet to the origin). In other words, Q is the radius of the largest sphere (centered at the origin) which is contained in CW

The convex hull (CW) can be calculated using different quality criteria, depending on the definition of $\|g\| = 1$:

- Minimizing the maximum finger force: The applied forces are individually and independently upper-bounded (equal to 1)

$$CW = ConvexHull(\bigoplus_{i=1}^{n}\{w_{i,1}, \ldots, w_{i,m}\}) \tag{6.17}$$

  where $\bigoplus$ is the Minkowski sum of the primitive wrenches $w_{n,m}$
- Minimizing the total finger force: The sum of the magnitude of the forces at the contact points is upper-bounded (equal to 1)

$$CW = ConvexHull(\bigcup_{i=1}^{n_c}\{w_{i,1}, \ldots, w_{i,m}\}) \tag{6.18}$$

  A comparison between both criteria can be found in [20]

Finding the closest facet to the origin means finding the most difficult wrench to the grasp to apply. Also, the wrench in the opposite direction is the most difficult external wrench for the grasp to resist. The measure indicates how efficient the grasp is at handling this worst case [21]

Units: Depends on the units of w. If ρ with units of [*length*] is used in Eq. (6.12), then this measure would have [*force*] units

Lower limit: 0

Upper limit: Non determined

Advantages: This is a geometric representation of the smallest perturbation wrench that breaks the grasp, independently of its direction. The force-closure condition can be verified, when this measure is greater than zero. The closer it is to 1, the more efficient the grasp is. It has been widely used by the robotics community [7, 19, 21]

Limitations: This measure depends on the choice of the origin of the reference system used to compute torques. It has no physical or mechanical interpretation

---

Volume of the convex hull [21]

Objective: Maximize to obtain an optimum grasp
Description: This measure calculates the volume of the boundary of the set of all possible wrenches acting on the object:

$$Q = Volume(CW) \tag{6.19}$$

where CW is defined by Eq. (6.17) or (6.18)
Units: Depends on the units of w. If ρ with units of [*length*] is used in Eq. (6.12), then this measure would have $[force]^6$ units
Lower limit: 0
Upper limit: Not determined
Advantages: It remains constant independently of the used torque reference system [21]. It can be calculated using the Qhull[1] algorithm
Limitations: None

Smallest maximum wrench to be resisted based on decoupled wrenches [12]

Objective: Maximize to obtain an optimum grasp
Description: This measure was created to avoid the problem of non-comparability of forces and moments in the wrench definition
It is defined as an optimality criterion based on decoupling wrenches into two subsets consisting of pure forces w_f or pure moments w_τ. Let Q_1 and Q_2 measure the grasp's ability to resist unit forces and unit torques respectively:

$$Q_1 = \min_{w_f \in CW} \|w_f\| \tag{6.20}$$

$$Q_2 = \min_{w_\tau \in CW} \|w_\tau\| \tag{6.21}$$

Then, the optimum grasp is the one which maximizes Q_2 among all grasps which maximize Q_1. Physically, this means seeking grasps which best resist forces in the grip plane and among these, considering the one which best resist moments perpendicular to the grip plane to be the optimum
Units: [*force* · *length*]
Lower limit: 0
Upper limit: Not determined
Advantages: Avoid the problem of non-comparability between forces and moments
Limitations: Difficult to calculate

---

[1] http://www.qhull.org/

Normal Grasping Force [22]

Objective: Minimize to obtain an optimum grasp

Description: This measure takes into account the magnitudes of the applied forces as indicative of the force efficiency in the grasp. It indicates how much passive forces the grasp can produce on the object to resist external disturbances. The normal grasping force is defined by:

$$f_n = \sum_{i=1}^{k} f_{normal_i} \tag{6.22}$$

where k is the number of fingers and f_{normal_i} is the normal component of the finger force. Then, for a given grasp and applied finger forces that resist a given external wrench w_{ext}, the quality of the grasp is given by:

$$Q = \frac{1}{f_n} \tag{6.23}$$

This measure should be minimized to obtain an optimum grasp

Units: $[1/force]$

Lower limit: 0

Upper limit: ∞

Advantages: The magnitude of the forces needs to be calculated

Limitations: It is the only measure found that takes the actual magnitude of the forces applied into account

---

6.2.1.4 Configuration of the Manipulator

The *manipulability* of a robot describes its ability to move freely in all directions in the workspace. Manipulability can measure either the ability to reach a certain position or the ability to change the position or orientation at a given configuration [23].

A *singular configuration* of a robot manipulator is a configuration at which the manipulator Jacobian (defined in Sect. 2.3.1) drops rank i.e. two or more of the columns of J become linearly dependent (see [23] for a thorough explanation). A singular configuration corresponds to a configuration in which the manipulator is not able to achieve instantaneous motion in certain directions (six degree of freedom manipulator) or the number of degrees of freedom of the end-effector drops (fewer than six degrees of freedom). Near singular configurations, the size of the joint velocities required to maintain a desired end-effector velocity in certain directions can be extremely large. Singularities also affect the size of the end-effector forces that the manipulator can apply. In order to avoid these difficulties, singular configurations should usually be avoided, if possible.

In addition to singularities of the Jacobian, a robot can also lose degrees of freedom when the joint variables are constrained to lie in a closed interval. In this case, a loss of freedom of motion can occur when one or more of the joints is at the limit of its travel. At such a configuration, motion past the joint limit is not allowed and the motion of the end-effector is restricted.

In order to study the local manipulability, there have been several measures proposed in the literature. Many of these measures rely on the singular values of the Jacobian J of the manipulator. Other penalize the hand joints that are at their maximum limits. They most relevant are explained here in detail.

Posture of hand finger joints [24]

Objective: Minimize to obtain an optimum grasp

Description: This index assesses the manipulation ability of the robot hand at the grasp position, measuring how far each joint is from its maximum limits. It is calculated as:

$$Q = 1/n_q \sum_{i=1}^{n_q} \left(\frac{y_i - a_i}{a_i - y_{iM}} \right)^2 \tag{6.24}$$

where n_q is the number of joints and a_i is the middle-range position calculated as:

$$a_i = (y_{iM} - y_{im})/2 \tag{6.25}$$

where y_{iM} and y_{im} are the maximum and minimum angle limits of the joint i. The grasp is optimal when all hand joints are at the middle-range position, having a quality measure of zero, and it goes to one when all its joints are at their maximum angle limits

Units: None

Lower limit: 0

Upper limit: 1

Advantages: It is easy to calculate and it is normalized between 0 and 1

Limitations: None

Minimum singular value of **J** [25]

Objective: Maximize to obtain an optimum grasp

Description: The minimum singular values of the Jacobian corresponds to the minimum workspace velocity that can be produced by a unit joint velocity vector [23]. The corresponding eigenvector gives the twist in which the motion of the end-effector is most limited

$$Q = \sigma_{min}(J) \tag{6.26}$$

Units: [*length*]

Lower limit: 0

Upper limit: Not determined
Advantages: It represents the minimum velocity transmission ratio, the maximum force transmission ratio and the best accuracy
Limitations: It only gives a worst case analysis

Determinant of **J** [25, 26]

Objective: Maximize to obtain an optimum grasp
Description: The determinant of the Jacobian measures the volume of the velocity ellipsoid generated by unit joint velocity vectors [23]. The determinant of the Jacobian is the product of the singular values of J:

$$Q = \sqrt{det(JJ^T)} = \sqrt{\lambda_1 \lambda_2 \dots \lambda_m} = \sigma_1 \sigma_2 \dots \sigma_m \quad (6.27)$$

where $\lambda_1 \geq \lambda_2 \geq \dots \geq \lambda_m$ are the eigenvalues of JJ^T and $\sigma_1 \geq \sigma_2 \geq \dots \geq \sigma_m$ are the singular values of the manipulator Jacobian

Geometrically, this manipulability is proportional to the volume of the manipulability ellipsoid whose principal axes have the same magnitudes as singular values of the Jacobian matrix [26]. It represents an average mobility over all direction at the end-effector and has an analytical expression as a function of the joint angles, but it depends on the scale of the manipulator. This measure favours a grasp that, given certain velocities in the finger joints, produces the largest velocities on the grasped object

Units: $[length]^m$, where $m = 2$ or 3 for planar or spatial manipulators respectively
Lower limit: 0 (when the Jacobian is singular)
Upper limit: Not determined
Advantages: Geometrical interpretation
Limitations: It is not normalized and has both scale and order dependency. The scale dependency prevents a fair comparison between a longer manipulator and a shorter manipulator, and the order dependency (m) makes it impossible to derive the physical meaning of the manipulability. In order to solve these problems, [26] proposed to modify this measure for the following:

$$Q = \sqrt[m]{det(JJ^T)}/f_M \quad (6.28)$$

where $m = 2$ or 3 and f_M is a function of dimensions $[length]^2$ such as the total length of the manipulator squared. This measure has no units: $[length]^2/[length]^2$

Inverse of the condition number of **J** [27, 26]

Objective: Maximize to obtain an optimum grasp

Description: The *condition number* of a matrix is defined as the ratio of its maximum singular value to its minimum singular value. For the Jacobian, the inverse condition number gives a measure of the sensitivity of the magnitude of the end-effector velocity to the direction of the joint velocity vector [23]. It provides a normalized measure of the previous one:

$$Q = \frac{\sigma_{min}}{\sigma_{max}} \tag{6.29}$$

where σ_{min} and σ_{max} are the smallest and largest singular values of the manipulator Jacobian

It is a dexterity measure that considers the capability of the hand to move an object in any direction with the same gain, which implies a good manipulation ability [4]. When the contribution of each joint velocity is the same in all components of the object velocity, the transformation between the velocity domain in the finger joints and the velocity domain of the object is uniform. When the condition number is equal to one, the columns of H are vectors orthogonal to each other and with the same module, indicating a uniform transformation and a grasp with the maximum quality. It has been used as a measure of proximity to singularity and as a measure of kinematic accuracy [26]

Units: None

Lower limit: 0

Upper limit: 1

Advantages: This measure is non-dimensional and thus independent of the scale of the manipulator. Also its optimal value is known to be one, therefore it is normalized

Limitations: It cannot be expressed analytically as a function of joints angles [26]

6.2.1.5 Group E: Task Oriented Measures

The measures defined previously are all unrelated to the tasks to be performed, therefore they are only useful in a limited set of applications like pick-and-place operations. For more complicated applications is not assure that using these quality measure will result in the optimal grasp. In such cases, the task requirements and knowledge of the environment need to be incorporated in the quality measure, which should favors grasps capable of generate wrenches in the object that are relevant to the task. The set of all wrenches that are expected to be applied on the object defines the *task wrench space* (TWS). The following measure takes the task information into account.

Task-oriented ellipsoid [8]

Objective: Maximize to obtain an optimum grasp

Description:	In this measure, the task is considered to be modeled by ellipsoids in the wrench space of the object, called task ellipsoids. The quality measure is the radius of the largest task ellipsoid that can be embedded in CW (defined in Sect. 6.2.1.3). As in the measure *Smallest maximum wrench to be resisted*, in order to guarantee a stable grasp, the origin must be contained in the interior; giving values greater than zero. Other alternatives of this measure describe the task as convex polytopes instead of ellipsoids [28, 29]
Units:	Depends on the units of w
Lower limit:	0
Upper limit:	Not determined
Advantages:	Considers the task information
Limitations:	The process of modelling a task by a task ellipsoid is complicated

6.2.1.6 Combined Measures

In order to give an overall assessment of a grasp, different quality measures should be combined. However, the different physical meaning of the grasp indices available, each one having different units and ranges, makes the merging a challenging process. There have been some attempts in robotics to combine some of these measures to create a global quality index [4].

The most simple method is to calculate the algebraic sum of the quality measures in a single global index, considering that all of them have to be either maximized or minimized. Boivin et al. [30] combined two criteria for 2D models: a criterion measuring the finger extension of their three-finger robotic hand and a force line criterion which takes into consideration the different friction coefficients for object surfaces. They are normalized dividing them with their maximum values and the quality measure is calculated summing them. Although this method is the most intuitive, it assumes that all measures have equal importance over the global quality of a grasp.

Chinellato et al. [14] presented a variation of this approach normalizing the outcome of ten measures, such that each have their ideal theoretical best value equal to 0. They were then summed to obtain the merged criteria. This work was extended by Chinellato et al. [15], where each measure was defined taking into account three parameters: the friction coefficient, the error in the position of fingers and an index to consider the object weight. The indices were grouped into three categories: shared features (S), features related with grasp configuration (G) and features related with hand configuration (C); proposing two global quality indices with the sum of measures:

$$Q_G = \sum_{i=1}^{4} Q_{S_i} + \sum_{i=1}^{2} Q_{G_i} \tag{6.30}$$

$$Q_C = \sum_{i=1}^{4} Q_{S_i} + \sum_{i=1}^{6} Q_{C_i} \tag{6.31}$$

These two global indices were used to rank several grasps, considering only planar objects. [6] used five normalized quality measures to create a global quality index. Each index was normalized such that a measure equal to one would represent the best grasp. The global index was calculated multiplying each index by a weight and then choosing their minimum:

$$Q_G = \min_{i=1}^{5} \{Q_i \cdot weight_i\} \tag{6.32}$$

More recently, [31] proposed a similar approach to calculate a global grasp quality index weighting three measures. The difference is that the index is evaluated integrating them over the time of performing the task for a series of n cooperative manipulators.

$$Q_G = \frac{1}{t_f - t_o} \int_{t_o}^{t_f} \left\{ W_1 \cdot C_N + \frac{W_2}{n} \sum_{i=1}^{n} \left(\frac{D_i}{D_{i,max}} \right) + \frac{W_3}{n} \sum_{i=1}^{n} \left(1 - \frac{P_i}{P_{i,max}} \right) \right\} dt \tag{6.33}$$

where t_o and t_f are the initial and final time, C_N takes into account the position of the contact points (Grasp Isotropy Index described by Eq. (6.3)), D_i kinetic characteristics of manipulators and the grasped object (Determinant of J described by Eq. (6.27)) and P_i evaluates the consumed power of the actuators considering the kinematic characteristics of the cooperative manipulators. Different weights W_1, W_2 and W_3 are included, whose sum is equal to one, to put different emphasis on each term.

Despite these attempts to create a global index, there are still a lot of open research problems in this area, such as which quality measures are needed to evaluate the different aspects of a grasp and how to combine them given their different units and ranges into an overall index assigning proper importance to each of its aspects.

6.2.2 Ergonomic Grasp Quality Measures

Ergonomics is concerned with the understanding of interactions among humans and other elements of a system. Grasp quality measures in ergonomics aim to measure the simulated postures adopted while grasping objects with different purposes. For example, study how the size and shape of objects affect the grasp according to the anthropometry of the different people that have to interact with them.

Usually, qualitative measures are used to assess the design of products which can be grasped with postures that are as close as possible to the relaxed posture of the human hand. However, this measure is equivalent to one of the previously presented robotic indicators (Sect. 6.2.1.4). In this section, we present two quantitative measures found in the literature that take into consideration ergonomic aspects such as the structure of the human hand/arm or the comfort when performing different grasps.

Easy of Grasping [3]

Objective: Maximize to obtain an optimum grasp

Description: This measure evaluates how the finger joint angles of the hand affect the subject feeling of ease of grasping an object

The method consists of the following steps:

1. **Pre-process**: Construct an "Ease of Grasping Evaluation Map" (EOG)

   It is created measuring the finger joint angles of a number of subjects with a dataglove when they were required to hold a set of objects including primitive shapes and real daily products. The subjects are asked to give a two level evaluation of the posture: "easy" or "not easy" to grasp. The recorded set of joints angles are processed using a Principal Component Analysis (PCA) and the results are plotted as points on the EOG evaluation map (an example is shown in Fig. 6.2) The large variance in human hand sizes was taken into account classifying the subjects into 5 different groups with respect of their hand dimensions and generating a EOG evaluation map for each group. After the map is created, it is possible to estimate the ease of grasping which is generated from a product model and a digital hand by plotting the principal component score for this posture on the map
2. An initial estimated grasp posture with the digital hand is generated for a product model to be grasped, and the principals component scores are calculated and plotted on the corresponding EOG evaluation map
3. The ease of grasping for the optimized grasp posture is obtained by the category assigned to the voxel that includes the evaluated posture on the map

Units: None

Lower limit: "not easy" to grasp

Upper limit: "easy" to grasp

Advantages: It is able to give an automatic ergonomic assessment of the easiness of grasping different objects

Limitations: Significant number of experiments need to be performed to create a EOG evaluation map. A single measure is perhaps not enough to evaluate the ergonomic quality of a grasp

---

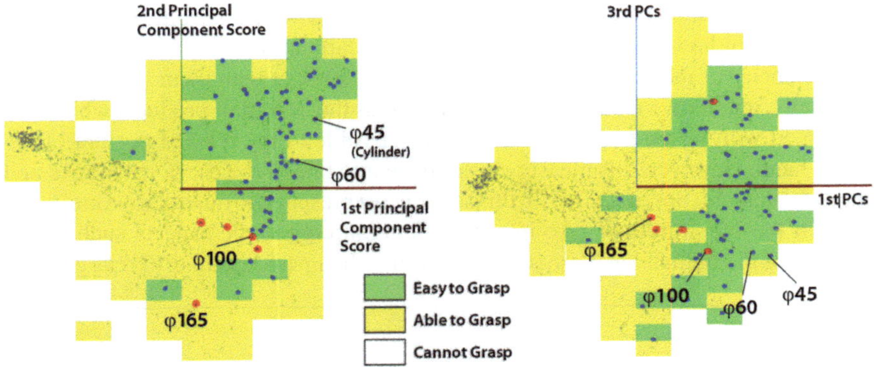

Fig. 6.2 Example of an "ease of grasping" evaluation map [32]

RULA: Rapid Upper Limb Assessment [33]

Objective: Minimize to obtain an optimum grasp

Description: This measure was developed to investigate the exposure of individual workers to risk factors associated with work-related upper limb disorders

The aim is to identify the muscular effort which is associated with a working posture, exerting force and performing static or repetitive work, and which may contribute to muscle fatigue

It has developed through the evaluation of the postures adopted; the number of movements and forces required; and muscle actions of operators working in a variety of manufacturing tasks. The body was divided into two groups: (A) the arm and wrist, and (B) the neck trunk an legs

$$\begin{aligned} \textit{RULA score} = & \ (\textit{Score A} + \textit{Muscle Use A} + \textit{Force A}) \\ & + (\textit{Score B} + \textit{Muscle Use B} + \textit{Force B}) \end{aligned} \tag{6.34}$$

Scores A and B are assigned based on a series of figures created for each segment and its range of movement, where 1 is given to postures where the risk factors are minimal and higher numbers indicate an increasing presence of risk factors

The procedure start by observing the operator during several work cycles in order to select the tasks and postures for assessment. Then using the figures the observer records the posture scores for both groups in an Employee Assessment Worksheet shown in Fig. 6.3

This measure has been used in robotics by [34] to enable a humanoid robot to produce human-like arm configurations which look more natural

Units: None

Lower limit: Score 1

Upper limit: Score 7
Advantages: It provides a rapid assessment of the loads on the musculoskeletal system of operators due to posture, muscle function and the forces they exert. It does not need any special equipment
Limitations: A person needs to be trained to use it in order to evaluate human postures, however it can be automatically calculated in a simulator. Since the human body is a complex and adaptative system, simple methods like RULA will not lead to unequivocal actions to eliminate any risk in the operators, then it can only be used as a guide

6.3 Adapted Robotic Grasp Quality Measures

We have selected eleven grasp quality measures and adapted them to evaluate the human grasp. From the indicators proposed in robotics, the ones that consider the algebraic properties of the matrix G are selected. Also, all the indicators that consider the distribution of the contact points excluding the ones that consider the Independent Contact Regions given that they are difficulty to calculate. From the measures that consider the magnitude of forces, we excluded the one that decouples the wrenches given that the smallest maximum wrench to be resisted and the volume of the convex hull are more commonly used and easier to calculate. Finally, two measures were selected that take into account the configuration of the manipulator: the posture of hand finger joints and the inverse of the condition number, which uses the Jacobian as the other measures reviewed in this group. Task measures are excluded at this point, considering that we have limited the experiments in this work to analyse static grasp holding the object for transport tasks. More complex task and measures that consider them are left for future work. Combined indexes are not used as they do not consider a general set of grasp measures and were not designed for the human hand, therefore we propose a combined index in the next chapter which is better-adapted to human grasp.

None of the measures from the ergonomics group were chosen. The Easy of grasping is difficult to obtain given that a significant number of experiments with subjective evaluations are needed to create the EOG evaluation map. The RULA was designed to assess the risk of the upper limb and it is not focus on the hand, therefore it will need to be adapted for the assessment of human grasp. Additionally, it needs specific information about the task such as the load, duration or repeatability which is not currently available, thus it should be considered also in future work.

We have divided the selected measures into four groups according to the information they use to evaluate the grasp: grasp matrix, distribution of contact points, magnitude of forces or manipulator configuration. In order to make them comparable, we propose how to normalize each measure (denoted with the subscript N) so that they have the best value of 1 and the worst value of 0.

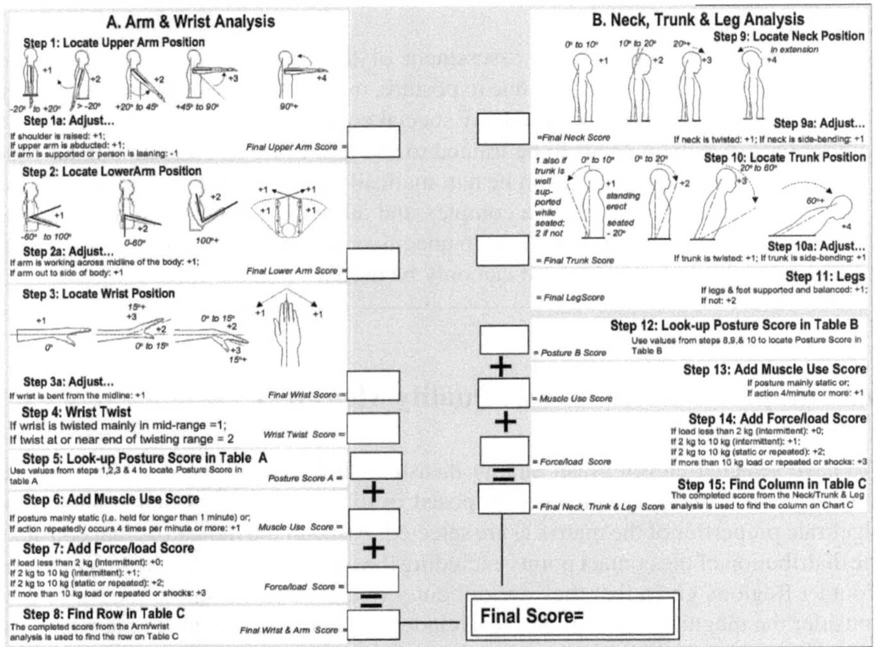

Fig. 6.3 RULA Employee assessment worksheet [33]

6.3.1 Group A: Algebraic Properties of G

They consider the algebraic properties of the grasp matrix G, and are defined to measure the grasp capability of withstanding external wrenches. The measures of this group can be directly applied to evaluate the human grasp.

6.3.1.1 Q_{A1}: Smallest Singular Value of G

It measures how far the grasp configuration is from falling into a singular configuration, losing the capability of withstanding external wrenches [8]. When a grasp is in a singular configuration, at least one of the singular values of G is zero. It is calculated as:

$$Q_{A1} = \sigma_{min}(G) \tag{6.35}$$

where $\sigma_{min}(G)$ is the smallest singular value of the matrix G. It has to be maximized and has no units. The lower limit is zero and the upper limit is not determined.

6.3.1.2 Q_{A2}: Volume of G in the Wrench Space

It gives an idea of the global contribution of all the contact forces [8], and can be calculated as:

$$Q_{A2} = v(G) = \prod_{i=1}^{r} \sigma_i \tag{6.36}$$

where r is the rank of G, and $\sigma_1 \geq \sigma_2 \geq \cdots \geq \sigma_r$ denote the nonzero singular values of G. This measure should be maximized and has no units. The lower limit is zero and the upper limit is not determined.

6.3.1.3 Q_{A3} : Grasp Isotropy Index

It looks for a uniform contribution of the contact forces to the total wrench exerted on the object, i.e. tries to obtain an isotropic grasp where the magnitudes of the internal forces are similar [6]. It is calculated as:

$$Q_{A3} = \frac{\sigma_{min}(G)}{\sigma_{max}(G)} \tag{6.37}$$

This measure has to be maximized and has no units. It approaches to one at a desirable configuration (isotropic) and is equal to zero at the singular configuration. Therefore, it is already normalized into the range 0 to 1. The normalized measure can be expressed as:

$$Q_{A3_N} = Q_{A3} \tag{6.38}$$

6.3.2 Group B: Distribution of Contact Points

These are indicators that use the distribution of the contact points. Better stability is assumed when contact points are distributed in a uniform way on the object surface and around the object centre of mass aiming to minimize the effect of gravitational and inertia forces.

6.3.2.1 Q_{B1}: Distance between the Centroid of the Contact Polygon and the Object's Center of Mass

It aims to minimize the effect of gravitational and inertia forces during the motion of the robot, measuring the distance between the centre of mass p of the grasped object and the centroid of the contact points p_c [10]. The centroid of the contact points is calculated as:

$$p_c = \frac{1}{n_c} \sum_{i=1}^{n_c} c_i \tag{6.39}$$

where n_c is the number of contact points and c_i is the location of each contact point. Then the measure is calculated as:

$$Q_{B1} = distance(p, p_c) \tag{6.40}$$

This measure has to be minimized and has units of length. We propose its normalization taking into account that its lower limit is zero and the upper limit can be calculated as the maximum distance from the centre of mass of the object to any point in the object's contour ($distance_{max}$), which can be obtained as the maximum distance from the center to any of the corners of the object bounding box. Additionally, the measure has been adapted to have 1 as its best value:

$$Q_{B1_N} = 1 - Q_{B1}/distance_{max} \tag{6.41}$$

6.3.2.2 Q_{B2}: Area of the Grasp Polygon

This measure is defined as the area of the polygon formed by the contact points. It has been used in robotics for three finger hands, where all finger points lie in a plane. In three-finger grasps, a larger triangle formed by the contact points on the object is argued to give a more robust grasp, i.e. with the same finger force, the grasp can resist larger external torques [13]. For the five fingers of the human hand, the indicator is extended using the method proposed by [35]. The contact plane is generated by selecting three fingers (thumb, index and middle fingers given their leading role in grasping). The remaining contacts are perpendicularly projected onto that plane (see Fig. 6.4). The index is calculated as:

$$Q_{B2} = Area(Polygon(p1, p2, p3, p4_P, p5_P)) \tag{6.42}$$

where $p1, p2, p3$ are the contact points for the thumb, index and middle fingers, and $p4_P, p5_P$ are the projected points of the ring and little fingers onto the plane.

This measure has to be maximized and has units of [*area*]. We propose to normalize this measure taking into account that its lower limit is zero and the upper limit ($Area_{max}$) can be calculated as the area of the polygon when the hand is open in a plane with the joints at their maximum abduction limits (Fig. 6.5). Then, the normalized measure can be calculated as: $Q_{B2N} = Q_{B2}/Area_{max}$.

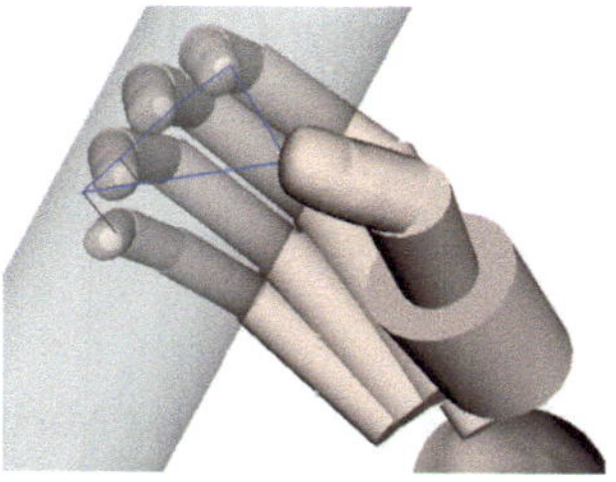

Fig. 6.4 Grasp polygon (*blue lines*) formed with the thumb, index and middle finger contact points and the projection (*black lines*) onto that plane of the ring and little finger contact points

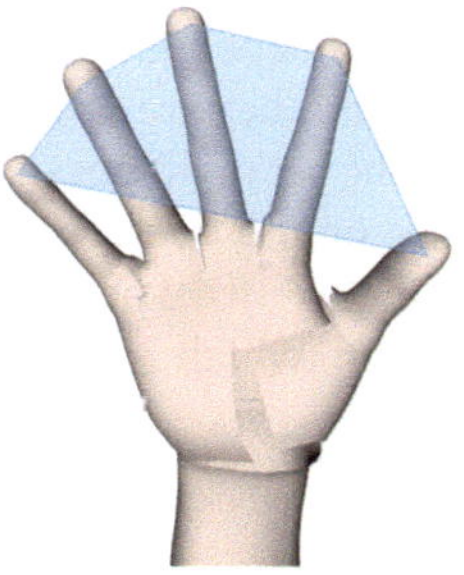

Fig. 6.5 Hand open posture to calculate the maximum contact polygon area

6.3.2.3 Q_{B3}: Shape of the Grasp Polygon

This measure is defined for planar grasp polygons and compares how far the internal angles of the grasp polygon are from those of the corresponding regular polygon [6]. For the five fingers of the human hand, a planar grasp polygon is obtained in the same way as for Q_{B2}. This index is calculated as:

$$Q_{B3} = \frac{1}{\theta_{max}} \sum_{i=1}^{n_f} |\theta_i - \bar{\theta}| \tag{6.43}$$

where n_f denotes the number of fingers, θ_i is the inner angle at the ith vertex of the polygon, $\bar{\theta}$ denotes the average angle of all inner angles of the grasp polygon and θ_{max} is the sum of the differences between the internal angles when the polygon has the most ill-conditioned shape (degenerates into a line) and those of the regular polygon (Eq. 6.9). This measure has to be minimized and has no units. The lower limit is zero and upper limit is 1. We propose to adapt this measure to have 1 as its best value:

$$Q_{B3_N} = 1 - Q_{B3}. \tag{6.44}$$

6.3.3 Group C: Magnitude of Forces

These are stability indicators that take into account the magnitude of forces applied at the contact points.

6.3.3.1 Q_{C1}: Smallest Maximum Wrench to be Resisted

The grasp quality is defined as the largest perturbation wrench that the grasp can resist with independence of its direction [5]. Only the directions of forces are used and their magnitudes are upper-bounded to 1. Defining *GWS* (see Sect. 2.3.2 Definition 2.10) as the set of all possible wrenches w acting on the object, the maximum of $w \in GWS$ lies on the boundary approximated as the convex hull over the discretized frictions cones (CW). Then the quality metric is the radius of the largest sphere centred at the origin, which is contained in GWS:

$$Q_{C1} = \min_{w \in CW} \|w\| \tag{6.45}$$

This measure has to be maximized and it has [*force*] units if the torque in w is divided by a parameter ρ with units of [*length*]. The index depends on the choice of the origin of the reference system used to compute torques. In this work, we use the centre of mass of the object and limited the magnitude of the torques to 1 choosing ρ as $distance_{max}$ defined previously for the measure Q_{B1}. Then, the upper limit of the index is $\sqrt{2}$ and the lower limit is zero. Then, we can normalize this index as:

$$Q_{C1_N} = \frac{Q_{C1}}{\sqrt{2}} \tag{6.46}$$

6.3.3.2 Q_{C2}: Volume of the Convex Hull

This measure is defined to avoid the dependence of the previous index on the selection of the origin of the reference system. The measure calculates the volume of the boundary of the set of all possible wrenches acting on the object [21].

$$Q_{C2} = Volume(CW) \tag{6.47}$$

The reference system and ρ have been chosen as described in the previous measure. The index has to be maximized and has units of $[force]^6$. Lower limit is zero and upper limit is no determined so that it is initially not possible to normalize the index in the range 0 to 1. However, we used the Monte Carlo method to estimate the upper limit.

Monte Carlo (MC) methods are stochastic techniques that use various distributions of random numbers to investigate problems. In this case, calculating the maximum

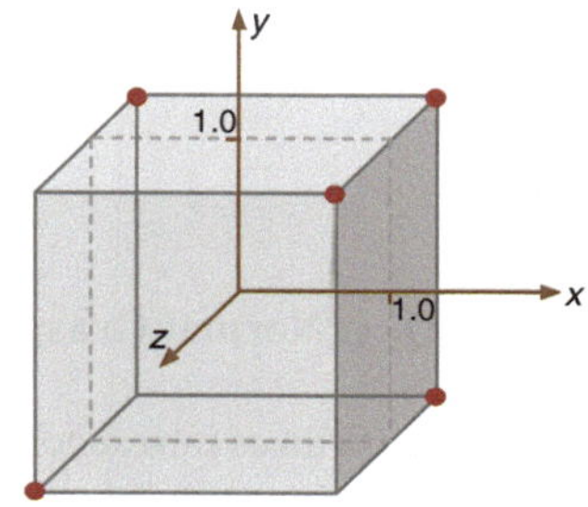

Fig. 6.6 Selected contact points

volume of the convex hull that can be generated by a grasp is a problem with no apparent solution. MC was used to randomly generate wrenches and determine Q_{C2} for a very large number of iterations. The variables that could be randomized are the contact normals $n_i = (n_{xi}, n_{yi}, n_{zi})$ and the contact points $c_i/\rho = (c_{xi}, c_{yi}, c_{zi})/\rho$. For each of their components, their values can vary in the range between $[-1, 1]$. In order to assure that n_i is normalized, it has to satisfy the following equation:

$$n_{yi}^2 + n_{zi}^2 = (1 - n_{xi}^2) \tag{6.48}$$

which can be interpreted as the equation of a circle with radius $\sqrt{1 - n_{xi}^2}$. Therefore n_{yi} and n_{zi} can be calculated as:

$$n_{yi} = \sqrt{1 - n_{xi}^2} cos(\theta) \tag{6.49}$$

$$n_{zi} = \sqrt{1 - n_{xi}^2} sin(\theta) \tag{6.50}$$

Giving random values to n_{xi} in the range between $[1, -1]$ and to θ in the range between $[0, 2\pi]$, we obtain a normalized value of n_i uniformly distributed.

In order to give values to c_i, initial experiments were performed given random values between $[-1, 1]$ to c_{xi}, c_{yi}, c_{zi}. They were represented as a cube centred at the origin and with dimensions $2 \times 2 \times 2$. As it was expected, greater volume is obtained when the contacts approach the surface of the cube. Therefore, in order to maximize the volume and minimize the time and computational costs, 5 contacts (corresponding to the 5 fingers of the human hand) in the cube boundary are chosen to assure that the maximum is obtained (see Fig. 6.6). The parameter ρ is calculated as $\sqrt{3}$ from the centre of the cube to one of the corners.

Having the contact points fixed, n_{xi} and θ are randomized using different number of iterations until two consecutive trials give the same results with an allowable error. The maximum and minimum volumes for each set of wrenches with the number of iterations performed are shown in Fig. 6.7.

The maximum value found after 40 million iterations was 0.7673 N^6 and therefore this is the value used as $Volume_{max}$ to normalize Q_{C2}. The normalized measure then can be calculated as:

$$Q_{C2_N} = Q_{C2}/Volume_{max} \tag{6.51}$$

6.3.3.3 Q_{C3}: Normal Grasping Force

This measure takes into account the magnitudes of the applied forces as indicative of the force efficiency in the grasp, because the magnitude of the perturbation wrench that the grasp can resist is related to the sum of the magnitudes of the contact normal forces:

$$Q_{C3} = \frac{1}{\sum_{i=1}^{k} f_{n_i}} \tag{6.52}$$

where k is the number of fingers and f_{n_i} is the normal component of the finger force. This index needs the estimation of the contact forces using the biomechanical model of the hand, as it is described in Sect. 5.5.7. This measure has to be minimized and has units of $[force]^{-1}$. Its lower limit is zero and upper limit tends to ∞.

6.3.4 Group D: Configuration of the Manipulator

The measures presented here are intended as manipulability robotic indices, describing the ability to reach a certain position or to change the position or orientation at a given configuration.

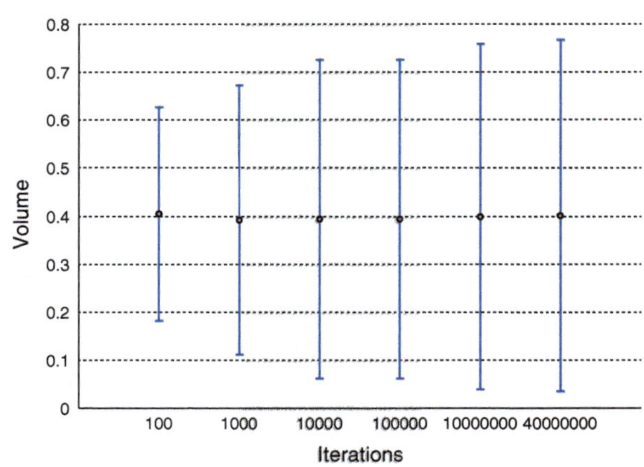

Fig. 6.7 Maximum and minimum volume calculated using MC method for different number of iterations

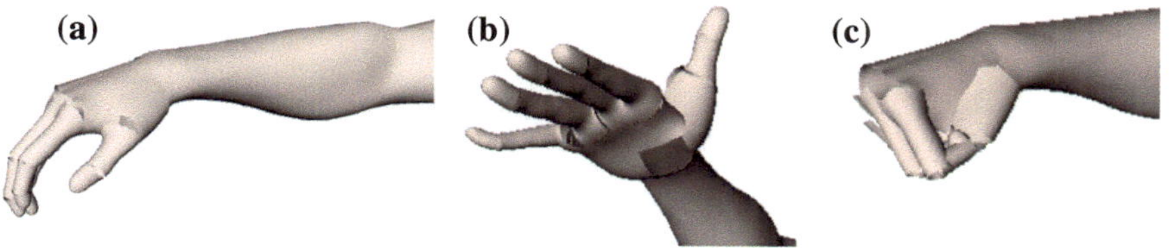

Fig. 6.8 Human hand postures: (**a**) Relaxed posture which is assigned the optimal quality measure; postures at their (**b**) minimum and (**c**) maximum joint limits corresponding to the grasps with the worst quality measures

6.3.4.1 Q$_{D1}$: Posture of Hand Finger Joints

This index measures how far each joint i is from its maximum limits [24]:

$$Q_{D1} = 1/n_q \sum_{i=1}^{n_q} \left(\frac{y_i - a_i}{R_i} \right)^2 \tag{6.53}$$

where n_q is the number of hand joints and R_i is the joint angle range between the middle-range position a_i and either the upper or lower angle limit, used to normalize the index. It is defined as:

$$R_i = \begin{cases} a_i - y_{im} & \text{if } y_i < a_i \\ y_{iM} - a_i & \text{if } y_i > a_i \end{cases}$$

where y_{iM} and y_{im} are the maximum and minimum angle limits of the ith joint (see Table 5.1).

The index has to be minimized, so that the grasp is optimal when all joints are at the middle-range position, having a quality measure of zero, and it goes to one when all its joints are at their maximum angle limits (Fig. 6.8). To adapt this index to the human hand, the joint angles defining the relaxed hand posture (see Sect. 5.5.1) have been used to define a_i. The measure has been modified to have 1 as its best value:

$$Q_{D1_N} = 1 - Q_{D1} \tag{6.54}$$

6.3.4.2 Q$_{D2}$: Inverse of the Condition Number of G$_J$

The condition number of a matrix is defined as the ratio of its maximum singular value to its minimum singular value. For the Jacobian, the inverse condition number gives a measure of the sensitivity of the magnitude of the end-effector velocity to the direction of the joint velocity vector. It is a dexterity measure that considers the capability of the hand to move an object in any direction with the same gain, which implies a good manipulation ability [4]:

$$Q_{D2} = \frac{\sigma_{min}(G_J)}{\sigma_{max}(G_J)}$$

where σ_{min} and σ_{max} are the smallest and largest singular values of the grasp Jacobian matrix G_J (defined in Sect. 2.3.1). This measure has to be maximized and has no units. Lower limit is zero and upper limit is one, indicating a uniform transformation and a grasp with the maximum quality. Therefore, this measure is already normalized into the desired range 0 to 1:

$$Q_{D2_N} = Q_{D2} \tag{6.55}$$

6.4 Biomechanic Grasp Quality Measures

In addition to the adapted robotic quality measures, we propose the use of two new biomechanical quality indicators.

6.4.0.3 Q_{E1}: Fatigue Index

This quality measure uses the common definition of fatigue proposed by [36], widely used in biomechanics, to measure the fatigue caused to the muscles when performing a grasp:

$$Q_{E1} = \sum_{i=1}^{m} \left(\frac{F_i}{PCSA_i} \right)^2 \tag{6.56}$$

where m represents the number of muscles, F_i the force exerted by each muscle and $PCSA_i$ its physiological cross-sectional area. The smaller the fatigue index the better will be the grasp. This index needs the estimation of the muscle forces from the use of a biomechanical model of the hand. It has units of $[force]^2 \times [area]^{-2}$. Its lower limit is 0 and its upper limit is the sum of the maximum stresses S_{max} the muscles can bear, which has been considered the same for all muscles [1]. We propose to normalize the index as:

$$Q_{E1_N} = 1 - \frac{Q_{E1}}{m \cdot {S_{max}}^2} \tag{6.57}$$

where the number of muscles (m) = 34 in our biomechanical model and $S_{max} = 35$ N/cm^2.

6.4.0.4 Q_{E2}: Muscle Safety Margin Index

This indicator measures the difference between the maximum force that can be exerted in the grasp posture with respect of the needed grasping force. The bigger the difference, the farther the hand is from its working limits and therefore, the better the quality of the grasp will be.

For a given grasp, the total normal force is calculated as the sum of the normal forces exerted by each finger:

$$F_n = \sum_{i=1}^{k} f_{n_i} \tag{6.58}$$

where k is the number of fingers. The problem of obtaining the normal forces (described in Sect. 5.5.7) consists in finding the appropriate set of contact forces that satisfy the dynamic equilibrium of the hand and the grasped object. This is an indeterminate problem that can be solved introducing a criterion chosen by the CNS with the minimization or maximization of an objective function. Therefore, to find the maximum force $F_{n_{max}}$ that can be exerted by a given grasp, the criterion chosen is to maximize the sum of the contact normal forces. In a similar way, to find the needed normal forces to perform the grasp $F_{n_{grasp}}$, the criterion is the one used by the biomechanical model, commonly minimizing the fatigue caused to the muscles when performing the grasp.

The quality measure is calculated with their difference:

$$Q_{E2} = F_{n_{max}} - F_{n_{grasp}} \tag{6.59}$$

This measure can be normalized dividing the difference by $F_{n_{max}}$:

$$Q_{E2_N} = \frac{F_{n_{max}} - F_{n_{grasp}}}{F_{n_{max}}} \tag{6.60}$$

A summary of the selected quality measures is presented in Table 6.1.

6.5 Conclusion

In this chapter, a review of the proposed grasp quality measures in robotics and ergonomics is presented. We adapted the most common robotic grasp quality measures to evaluate the human grasp. The *fatigue index* and *the muscle safety margin index* were proposed to consider the biomechanical aspects of the human hand not taken into account by the existing robotic measures. More research is needed to investigate other biomechanical measures that might be obtained from the use of existing biomechanical hand models.

Table 6.1 Summary of selected quality measures

Name		Formula	Normalization	Units	Min	Max
Group A: Algebraic properties of G						
Q_{A1}	Smallest singular value of G	$\sigma_{min}(G)$	–	none	0	–
Q_{A2}	Volume of G in the wrench space	$\prod_{i=1}^{r} \sigma_i$	–	none	0	–
Q_{A3}	Grasp Isotropy Index	$\sigma_{min}(G)/\sigma_{max}(G)$	Q_{A3}	none	0	1
Group B: Distribution of contact points						
Q_{B1}	Distance between the centroid of the contact polygon and the object's centerof mass	$distance(p, p_c)$	$1 - Q_{B1}/distance_{max}$	none	0	1
Q_{B2}	Area of the grasp polygon	$Area(Polygon\ (p1, p2, p3, p4_P, p5_P))$	$Q_{B2}/Area_{max}$	none	0	1
Q_{B3}	Shape of the grasp polygon	$\frac{1}{\theta_{max}} \sum_{i=1}^{n_f} \lvert\theta_i - \bar{\theta}\rvert$	$1 - Q_{B3}$	none	0	1
Group C: Algebraic properties of G						
Q_{C1}	Smallest maximum wrench to be resisted	$\min_{w \in CW} \lVert w \rVert$	$Q_{C1}/\sqrt{2}$	[*force*]	0	1
Q_{C2}	Volume of the convex hull	$Volume(CW)$	$Q_{C2}/Volume_{max}$	none	0	1
Q^*_{C3}	Normal Grasping Force	$1/\sum_{i=1}^{k} f_{n_i}$	–	[1/*force*]	0	∞
Group D: Configuration of the manipulator						
Q_{D1}	Posture of hand finger joints	$1/n_q \sum_{i=1}^{n_q} ((y_i - a_i)/R_i)^2$	$1 - Q_{D1}$	none	0	1
Q_{D2}	Inverse of the condition number of $\mathbf{G_J}$	$\sigma_{min}(G_J)/\sigma_{max}(G_J)$	Q_{D2}	none	0	1
Group E: Biomechanic Grasp Quality Measures						
Q_{E1}	Fatigue Index	$\sum_{i=1}^{m} (F_i/PCSA_i)^2$	$1 - (Q_{E1}/m \cdot {S_{max}}^2)$	none	0	1
Q_{E2}	Muscle safety margin index	$F_{n_{max}} - F_{n_{grasp}}$	$Q_{E2}/F_{n_{max}}$	none	0	1

*Measure Q_{C3} is not normalized and it is the only measure that has to be minimized to produce the best grasp

The measures were normalized, when possible, in the range between 0 and 1. The Monte Carlo method has been used to successfully obtain the upper limit for the robotic measure Q_{C2} in its adaptation of human grasp evaluation in order to illustrate its potential for the normalization of any of the measures. Moreover, the ranges of variation which are better-adapted to real human grasping might be also achieved through its use.

In the next chapters, these measures are used to evaluate the quality of different grasps. Analysis of variability and sensitivity are performed with the aim of finding the minimum necessary measures able to evaluate the different aspects of a grasp and propose how to combine them to obtain an overall quality index.

References

1. Sancho-Bru, J.L., Mora, M.C., León, B.E., Antonio Pérez-González, Iserte, J.L., Morales, A.: Grasp modelling with a biomechanical model of the hand. Computer Methods in Biomechanics and Biomedical Engineering, 1–14 (2012). http://www.tandfonline.com/doi/abs/10.1080/10255842.2012.682156
2. Goussous, F.A.: Grasp planning for digital humans. Ph.D. thesis, Iowa University (2007)
3. Endo, Y., Kanai, S., Kishinami, T., Miyata, N., Kouchi, M., Mochimaru, M.: A computer-aided ergonomic assessment and product design system using digital hands. In: Proceedings of the 1st international conference on Digital human modeling. pp. 833–842. Springer, Germany (2007)
4. Suárez, R., Roa, M., Cornella, J.: Grasp quality measures. Technical University of Catalonia, Tech. rep (2006)
5. Ferrari, C., Canny, J.: Planning optimal grasps. In: Proceedings 1992 IEEE International Conference on Robotics and Automation, pp. 2290–2295 (1992)
6. Kim, B.H., Oh, S.R., Yi, B.J., Suh, I.H.: Optimal grasping based on non-dimensionalized performance indices. In: Proceedings of the IEEE International Conference on Intelligent Robots and Systems. vol. 2, pp. 949–956 (2001)
7. Roa Garzón, M.: Grasp planning methodology for 3D arbitrary shaped objects. Ph. d. thesis, Universidad Politécnica de Cataluña (2009)
8. Li, Z., Sastry, S.: Task-oriented optimal grasping by multifingered robot hands. IEEE J. Robot. Autom. **4**(1), 32–44 (1987)
9. Ben-Israel, A.: A volume associated with m x n matrices. Linear algebra and its applications 167(0), 87–111 (1992). http://www.sciencedirect.com/science/article/pii/002437959290340G
10. Ding, D., Lee, Y.H., Wang, S.: Computation of 3-d form-closure grasps. IEEE Trans. Robot. Autom. **17**(4), 515–522 (2001)
11. Ponce, J., Sullivan, S., Sudsang, A., Boissonnat, J.D., Merlet, J.P.: On computing four-finger equilibrium and force-closure grasps of polyhedral objects. Int. J. Robot. Res. **16**(1), 11–35 (1997)
12. Mirtich, B., Canny, J.: Easily computable optimum grasps in 2-d and 3-d. In: Proceedings IEEE International Conference on Robotics and Automation. pp. 739–747 (1994)
13. Xiong, C., Li, Y., Ding, H., Xiong, Y.L.: On the dynamic stability of grasping. I. J. Robot. Res. 18(9), 951–958 (1999). http://dblp.uni-trier.de/db/journals/ijrr/ijrr18.html
14. Chinellato, E., Fisher, R., Morales, A., del Pobil, A.: Ranking planar grasp configurations for a three-finger hand. In: Proceedings of the IEEE International Conference on Robotics and Automation. vol. 1, pp. 1133–1138 (2003)
15. Chinellato, E., Morales, A., Fisher, R., del Pobil, A.: Visual quality measures for characterizing planar robot grasps. IEEE Trans. Syst., Man, Cybern., C: Appl. Rev. 35(1), 30–41 (2005)

16. Park, Y.C., Starr, G.P.: Grasp synthesis of polygonal objects using a three-fingered robot hand. Int. J. Robot. Res. **11**(3), 163–184 (1992)
17. Ponce, J., Faverjon, B.: On computing three-finger force-closure grasps of polygonal objects. IEEE Trans. Robot. Autom. **11**(6), 868–881 (1995)
18. Kirkpatrick, D.G., Mishra, B., Yap, C.K.: Quantitative steinitz's theorems with applications to multifingered grasping. In: Proceedings of the twenty-second annual ACM symposium on Theory of computing. pp. 341–351. STOC '90, ACM, NY, USA (1990)
19. Borst, C., Fischer, M., Hirzinger, G.: Grasping the dice by dicing the grasp. In: Proceedings IEEE/RSJ International Conference on Intelligent Robots and Systems (IROS 2003). vol. 4, pp. 3692–3697 (2003)
20. Zheng, Y., Qian, W.H.: Limiting and minimizing the contact forces in multifingered grasping. Mech. Mach. Theor. 41(10), 1243–1257 (2006). http://www.sciencedirect.com/science/article/pii/S0094114X05001953
21. Miller, A.T., Allen, P.K.: Examples of 3D grasp quality computations. In: Proceedingsof the IEEE International Conference on Robotics and Automation, vol. 2, pp. 1240–1246 (1999)
22. Liu, G., Li, Z.: Real-time grasping-force optimization for multifingered manipulation: theory and experiments. IEEE/ASME Trans. Mechatron. **9**(1), 65–77 (2004)
23. Murray, R.N., Li, Z., Sastry, S.: A mathematical introduction to robotics manipulation. CRC Press, US (1994)
24. Liegeois, A.: Automatic supervisory control of the configuration and behavior of multibody mechanisms. IEEE Trans. Syst. Man, Cybernet. **7**(12), 842–868 (1977)
25. Klein, C.A., Blaho, B.E.: Dexterity measures for the design and control of kinematically redundant manipulators. Int. J. Robot. Res. **6**(2), 72–83 (1987)
26. Kim, J.O., Khosla, P.: Dexterity measures for design and control of manipulators. Proc. IROS Workshop Intell. Robot. Syst. **91**, 758–763 (1991)
27. Salisbury, J.K., Craig, J.J.: Articulated hands: Force control and kinematic issues. Int. J. Robot. Res. **1**(1), 4–17 (1982)
28. Zhu, X., Ding, H., Li, H.: A quantitative measure for multi-fingered grasps. Proceedings IEEE/ASME International Conference on Advanced Intelligent. Mechatronics, In (2001). (1), 213–219 (2001)
29. Haschke, R., Steil, J., Steuwer, I., Ritter, H.: Task-oriented quality measures for dextrous grasping. In: Proceedings IEEE International Symposium on Computational Intelligence in Robotics and Automation. pp. 689–694 (2005)
30. Boivin, E., Sharf, I., Doyon, M.: Optimum grasp of planar and revolute objects with gripper geometry constraints. In: Proceedings of the IEEE International Conference on Robotics and Automation. vol. 1, pp. 326–332 (2004)
31. Cheraghpour, F., Moosavian, S., Nahvi, A.: Multiple aspect grasp performance index for cooperative object manipulation tasks. In: IEEE/ASME International Conference on Advanced Intelligent Mechatronics. AIM 2009. pp. 386–391 (2009)
32. Endo, Y., Kanai, S., Kishinami, T., Miyata, N., Kouchi, M., Mochimaru, M.: Virtual grasping assessment using 3d digital hand model. In: 10th Annual Applied Ergonomics Conference: Celebrating the Past—Shaping the Future (2007)
33. McAtamney, L.: Nigel Corlett, E.: RULA: A survey method for the investigation of work-related upper limb disorders. Appl. Ergon. **24**(2), 91–9 (1993)
34. Zacharias, F., Schlette, C., Schmidt, F., Borst, C., Rossmann, J., Hirzinger, G.: Making planned paths look more humanlike in humanoid robot manipulation planning. IEEE International Conference on Robotics and Automation (ICRA), In (2011)
35. Supuk, T., Kodek, T., Bajd, T.: Estimation of hand preshaping during human grasping. Med. Eng. Phys. **27**(9), 790–7 (2005)
36. Crowninshield, R.D., Brand, R.A.: A physiologically based criterion of muscle force prediction in locomotion. J. Biomech. **14**(11), 793–801 (1981)

Chapter 7
Human Grasping Simulation

7.1 Introduction

In the previous chapter, we proposed the adaptation of the most common robotic grasp quality measures to the evaluation of human hand grasps together with new complementary quality indices considering biomechanical aspects not taken into account by the actual robotic indices (summarized in Table 6.1). However, it is not efficient to calculate all reported quality measures for evaluating a grasp when performing several experiments and to analyze the high amount of data produced. To overcome this problem, a set of measures may be chosen as it is expected that some of them provide similar information as they were formulated to evaluate the same aspect.

There have been some attempts in robotics to combine some of the measures to create global quality indices (see Sect. 6.2.1.6). This was done by using the sum (or weighted sum) of a set of selected quality measures in a single global index, considering that all of them have to be either maximized or minimized [1–4]. These works showed the problem of merging indices with different numerical ranges and physical units. Each proposal used a different set of quality measures to create the global quality index. To our knowledge, there is not an accepted unique global index in the literature capable of measuring all aspects of robotic grasps.

One of the purposes of this chapter is to find the minimum set of indices that allows the evaluation of the different aspects of the human grasp on simulation. In order to do that, we had to make the indicators comparable given that they have different dimensional units and ranges of variation. As a first attempt, their mathematical limits were used to normalize them—in the cases that it was possible to determine these limits. In order to find more realistic ranges, we performed a variability analysis. A series of experiments were simulated, with different grasps of cylinders. We varied different parameters of the experiments such as the size or weight of the object or the number of fingers used to perform the grasp. This variability analysis served to propose new limits to normalize each measure.

B. León et al., *From Robot to Human Grasping Simulation*,
Cognitive Systems Monographs 19, DOI: 10.1007/978-3-319-01833-1_7,

The robustness of the indices was checked through the results of a sensitivity analysis of the indicators to the input parameters. This was done through a set of simulations obtained by randomly varying the input parameters of the grasps. This sensitivity analysis helped us in the appropriate selection of robust indices when trying to reduce the set of indicators. We also studied the correlations between the indicators, in order to find the minimum set of indicators allowing the evaluation of the different aspects of the human grasp. This allowed us to identify the independent aspects that are being measured by all the indices that have been calculated. We have provided a physical interpretation of these independent aspects, and proposed a global grasp quality index through the use of neural networks.

Finally in the last section, we present a possible application of the developed human grasp evaluation framework comparing the grasp capabilities of a prosthetic hand using the grasp quality measures with the ones obtained for the human hand using our model. This could give insights as how to improve the design of hand prosthesis or robotic hands by means of obtaining better quality scores.

7.2 Evaluation of Human Grasps

A set of experiments was designed to explore a wide range of possibilities of the human grasp enabling the analysis of the variability and sensitivity of the measures.

Nylon cylinders, 200 mm in length and with a 50 mm diameter, were chosen for the experiments (Fig. 7.1). Only grasps involving the distal phalanxes of the fingers and thumb were considered, taking into account that the frequency of grasping objects with the finger's distal phalanxes while performing common daily activities has been found to be three times the frequency of grasping objects with contacts along the fingers and the palm [5].

Three different grasps were selected: cylindrical, claw and lateral (Fig. 7.2). The cylindrical and lateral grasps were selected as the most intuitive, and were complemented with the claw grasp as a forced posture, which would only be performed in rare cases, especially by persons with some pathology.

Fig. 7.1 Nylon cylinder with 50 mm diameter used for the experiments

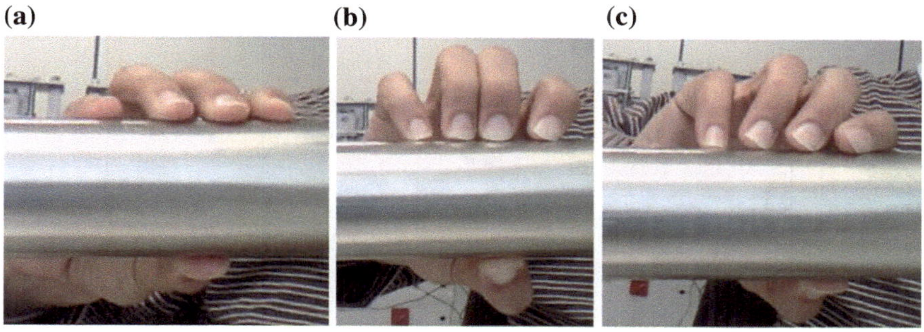

Fig. 7.2 The three types of grasps selected. **a** Cylindrical, **b** claw, **c** lateral

In order to vary the number of contact points, all grasps were performed using five and three fingers (thumb, index and middle fingers). The cylinder was grasped with these six grasps in vertical orientation at its centre and at one of its extremes. Joint angle data for the simulation of the postures were obtained from the registration of the 3D position of 29 markers on the hand of a healthy female subject using a VICON motion tracking system [6].

For the first group of experiments, a cylinder was used with a weight of 460 g. A total of 36 grasps were simulated (Fig. 7.3) and the results of the quality measures are presented in Fig. 7.4. *Y*-axes in this figure have different scales in order to better display the variation between values for each measure

Group A measures (see Table 6.1 for reference) present a similar behaviour, specially Q_{A1_N} and Q_{A3_N}. Lateral grasps were rated better than claw and cylinder grasps in most cases. In addition, horizontal grasps were rated better than the vertical ones. Q_{B1_N} rated grasp better at the centre than at the object's edges, as expected. Q_{B2_N} and Q_{B3_N} showed an inverse behavior given that grasps with three fingers have a grasp polygon with less area but closer to a regular triangle than the five finger grasps. Cylindrical grasps have more regular grasp triangles than claw grasps, which in turn have more than lateral grasps. However, lateral grasps have a more regular pentagon than cylindrical and claw grasps. Group C measures, Q_{C1_N} and Q_{C2_N} rated better 5-finger grasps than the 3-finger ones and, in most cases, the cylindrical grasps than the claw and lateral grasps.

Q_{C3} is the only measure that needs to be minimized to obtain a better grasp, as it could not be normalized. It rated better 3 than 5-finger grasps showing that bigger normal forces are applied to oppose potential object slides. In many cases, the biomechanical model did not find a feasible solution, specially for lateral grasps, for which the model should be improved to account for the passive articular forces that can counteract abduction-adduction moments. Horizontal grasps applied less normal forces than the vertical ones, therefore they were worst rated. Similarly, center grasps were rated worst than those grasping the object at its extremes.

Measure Q_{D1_N} gave high values (from 0.57 to 0.87) showing that the selected grasps are far from the hand joints' limits. A surprising result is that, in most cases,

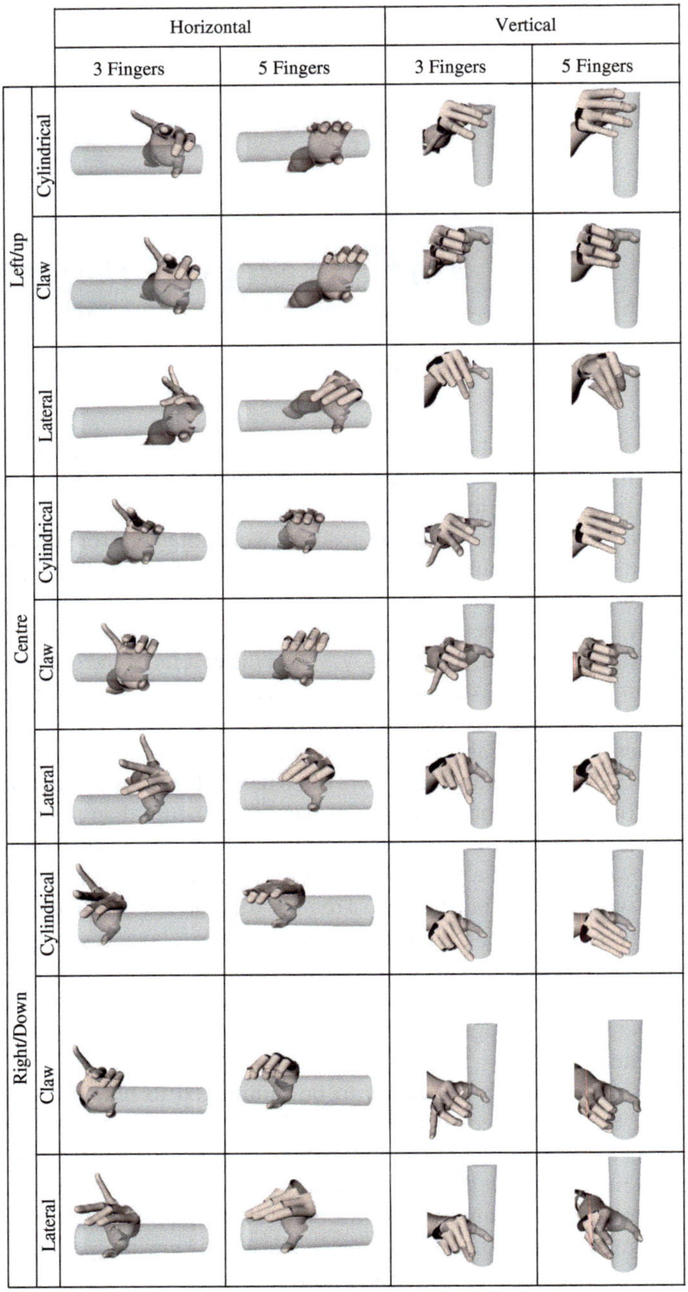

Fig. 7.3 36 selected grasp postures (cylinder: 50 mm diameter, 200 mm length and 460 g)

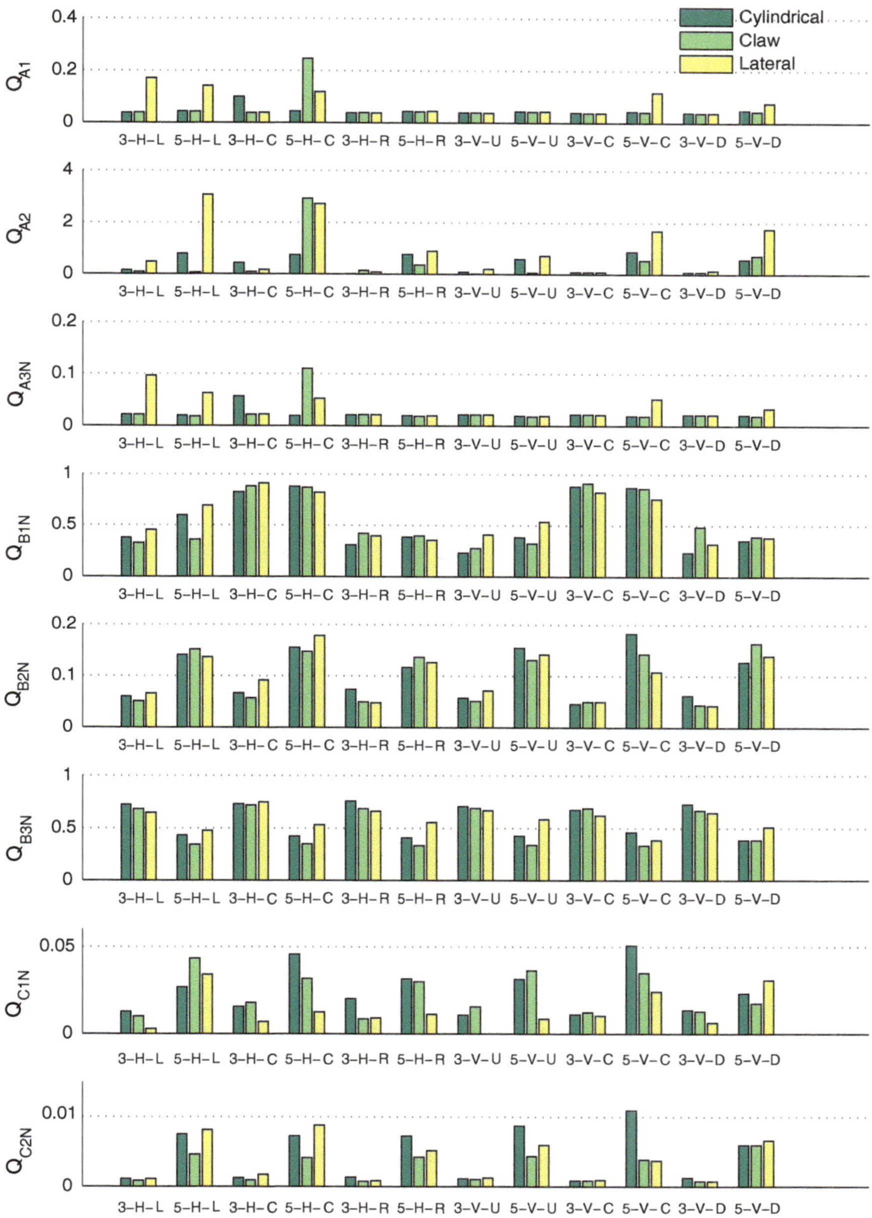

Fig. 7.4 Quality measures evaluated for 36 simulated grasps. Postures described by: number of fingers (3, 5)—orientation (*H* Horizontal, *V* Vertical)—position (*L* Left, *C* Center, *R* Right, *U* Up, *D* Down) (Cont.)

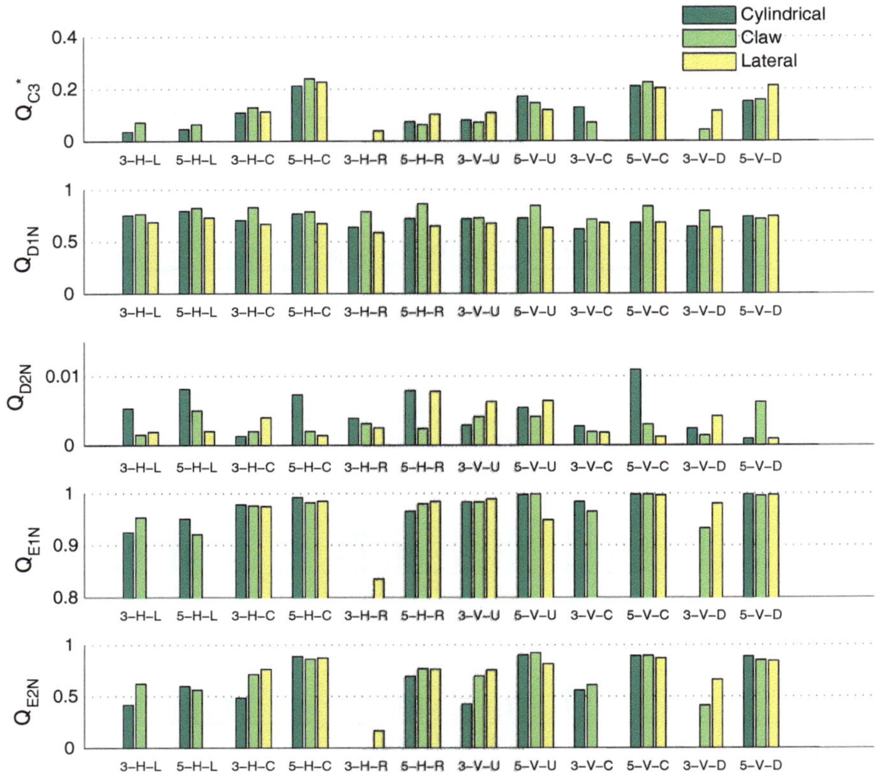

Fig. 7.4 (Continued)

claw grasps were better rated than cylindrical and lateral grasps. Claw grasps are in theory the most forced posture simulated but the measure does not reflect that. This is due to the fact that, although the distal and proximal interphalangeal joints are closer to their limits, the other joints are closer to the relaxed posture. This measure can be improved in different ways. On the one hand, different weights could be assigned to each joint. On the other hand, a correlation between the DIP and PIP joint flexion angles is expected, as reported in the literature [7]. Therefore, the measure could be modified to penalise the differences in the relationship between the flexion at DIP and PIP joints with respect to the relationship measured at the relaxed posture.

Conversely, Q_{D2_N} assigned most values close to zero (from 0.0009 to 0.0109) to the selected grasps. As it is a measure of the grasp manipulability, cylindrical and 5 finger grasps were better rated, instead claw grasps were rated worse.

Biomechanical measures, Q_{E1_N} and Q_{E2_N}, also did not find feasible solutions for some grasps, lateral ones in most cases. Values of Q_{E1_N} were very close to one (from 0.83 to 0.999), showing that the muscle stresses are far from their limits for the selected grasps. Grasping the cylinder with 3 fingers, horizontally and from the right

Table 7.1 Minimum and maximum values per measure over all postures. The posture giving the extreme value is also shown

	Q_{A1}	Q_{A2}	Q_{A3N}	Q_{B1N}	Q_{B2N}	Q_{B3N}
Min	0.0374 	0.0140 	0.0181 	0.2309 	0.0428 	0.3333 
Max	0.2462 	3.0704 	0.1101 	0.9122 	0.1822 	0.7605 

	Q_{C1N}	Q_{C2N}	Q_{C3}*	Q_{D1N}	Q_{D2N}	Q_{E1_N}	Q_{E2_N}
Min	0.0000 	0.0007 	0.0353 	0.5868 	0.0009 	0.8355 	0.1680 
Min	0.0506 	0.0109 	0.2401 	0.8622 	0.0109 	0.9984 	0.9183 

* Measure Q_{C3} is not normalized and has to be minimized to produce the best grasp

appeared as a especially difficult grasp for all the biomechanical measures together with Q_{C3}. Q_{E2_N} rated worse grasps with 3 than 5-fingers showing that they need to apply higher normal forces and therefore, they are closer to the force limits the grasp can exert.

The maximum and minimum values for each measure were identified and are shown in Table 7.1 with the corresponding grasp. Although most of the measures have been normalized, they seem to move in different zones within the range 0–1, in a non homogeneous distribution.

From these results, it seems that the proposed limits for some of the measures are not realistic, as they have been found to vary within very small ranges. Therefore, a variability analysis performing more grasps is presented in the next section in an attempt to estimate these limits more precisely.

7.3 Variability Analysis

Different experiments were designed to vary different aspects influencing the grasp to identify ranges of variation which are better-adapted to human grasping. This would help to find more realistic ranges of variation of the measures than those obtained with the mathematical limits, enabling the improvement of the normalization of these indicators.

7.3.1 Material and Methods

Three important aspects to vary were considered: the object's weight and size and the fingers used to perform the grasp.

7.3.1.1 Varying Object's Weights

Only three of the measures take the weight of the object into account: the Normal Grasping Force (Q_{C3}), the Fatigue Index (Q_{E1_N}) and the Muscle Safety Margin Index (Q_{E2_N}). For these measures, a lighter cylinder (180 g) was also used to perform the same grasps selected for the initial experiment (Fig. 7.3).

7.3.1.2 Varying Object's Sizes

Two additional cylinders of different diameters (25 and 75 mm) were also evaluated for the vertical grasps shown in Fig. 7.5.

7.3.1.3 Varying the Number of Fingers

Grasping the cylinders with different fingers was also considered. Three, four and five fingers were selected to grasp the medium cylinder (50 mm) shown in Fig. 7.6.

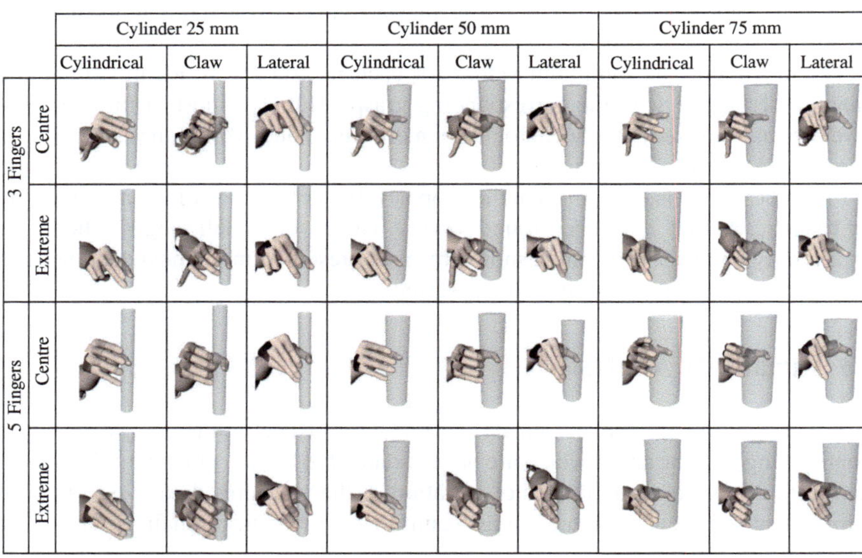

Fig. 7.5 Selected postures varying the cylinder's diameter

The quality measures were evaluated for the simulated grasps finding the maximum and minimum values for each measure.

7.3.2 Results

7.3.2.1 Varying Weights

Measures Q_{C3}, Q_{E1_N} and Q_{E2_N} are compared when grasping the light and the heavy cylinder in Fig. 7.7 and the minimum and maximum values were identified in Table 7.2.

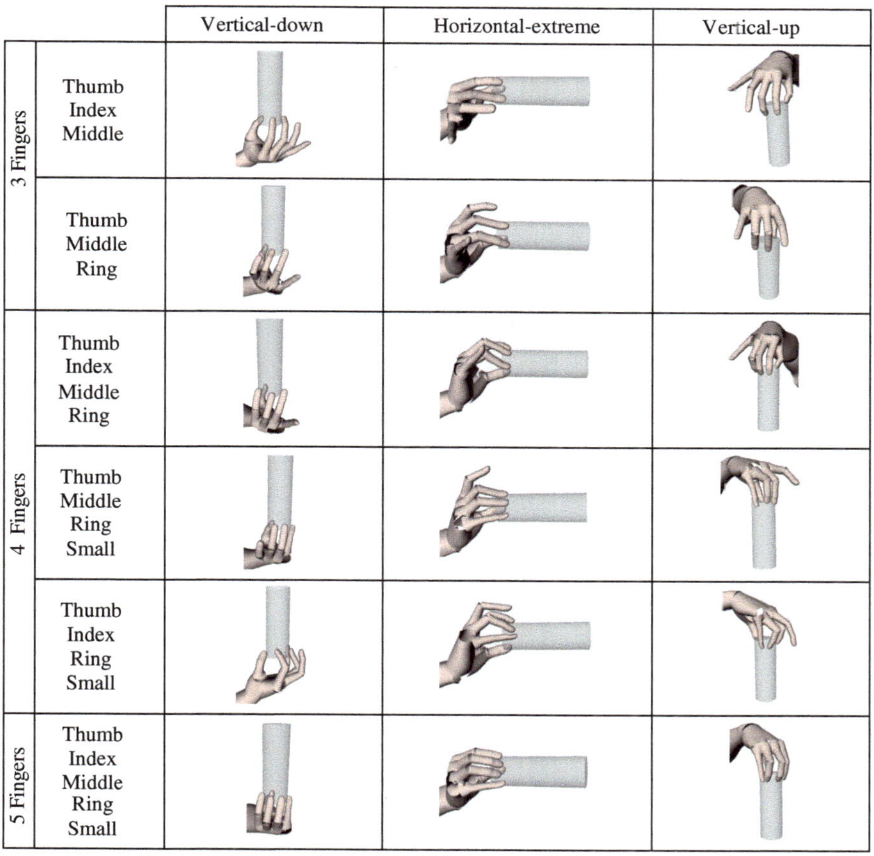

Fig. 7.6 Selected postures varying the fingers used in the grasp and the cylinder orientation

Q_{E1_N} and Q_{E2_N} provided the best results for the lighter cylinder, which is in accordance with the fact that grasping heavier objects results in a more fatiguing task. Q_{C3}, by contrast, always provided the best results for the heaviest cylinder. This is due to the fact that higher normal forces are required to grasp a heavier object, so that additional perturbation wrenches will not significantly affect the stability of the grasp.

Again, there were some lateral grasps for which the biomechanical model could not find a feasible solution (see Table 7.2). For Q_{E1_N} and Q_{E2_N}, the worst grasps were all horizontals and at the right, and the best were always the vertical 5-finger-claw grasp at the top. The biomechanical model does not take into account extreme stresses produced in the ligaments or joints, therefore it is not unreasonable that it

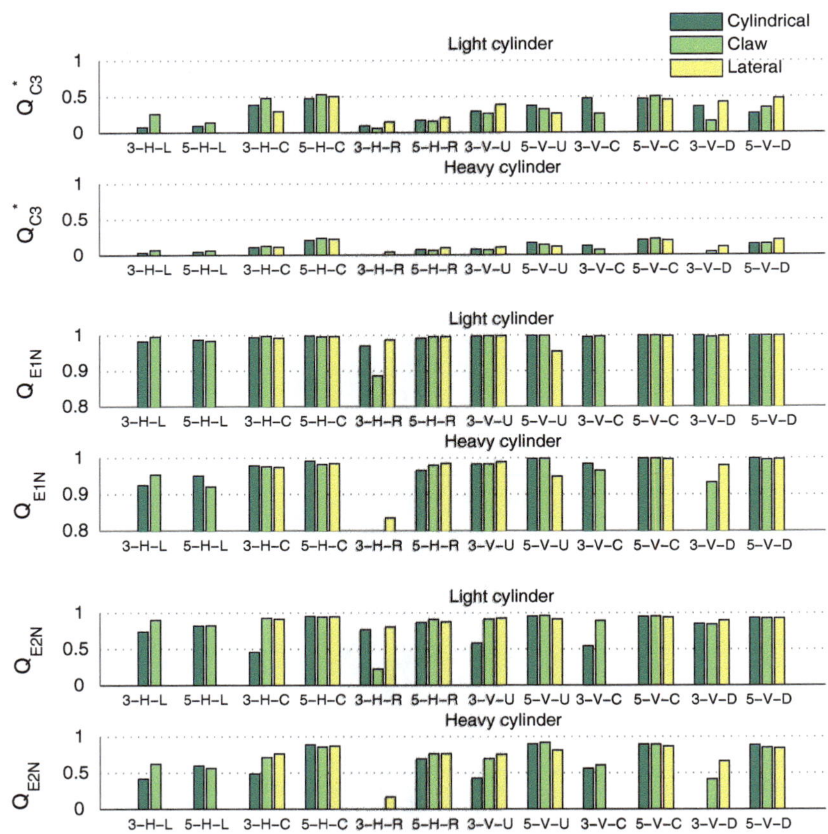

Fig. 7.7 Quality measures Q_{C3}, Q_{E1_N} and Q_{E2_N} evaluated for 36 simulated grasps for the light cylinder and for the heavy cylinder. Postures described by: number of fingers (3, 5)—orientation (*H* Horizontal, *V* Vertical)—position (*L* Left, *C* Center, *R* Right, *U* Up, *D* Down)

Table 7.2 Posture with the minimum and maximum grasp quality value for each measure when varying the object's weight

	Q_{C3}		Q_{E1_N}		Q_{E2_N}	
	Light Cylinder	Heavy cylinder	Light Cylinder	Heavy cylinder	Light Cylinder	Heavy cylinder
Not feasible	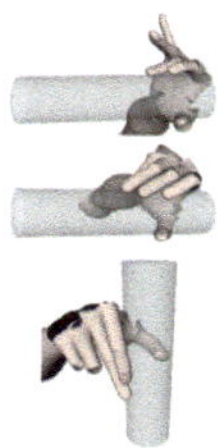	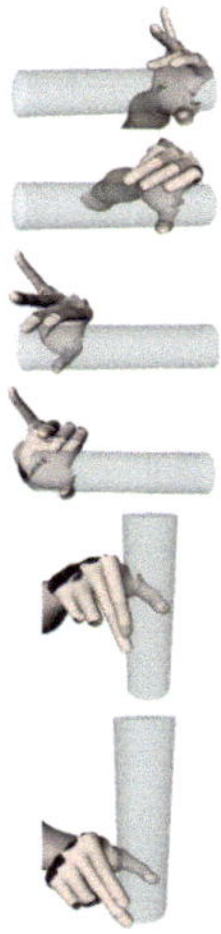	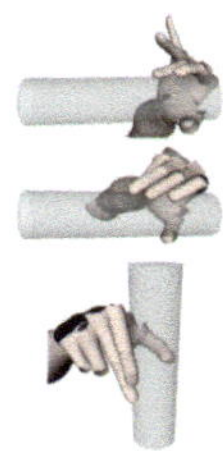	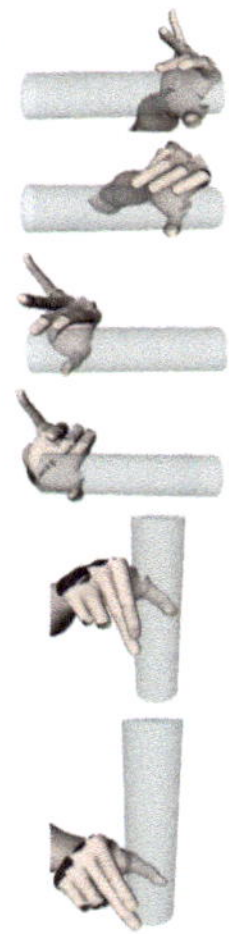	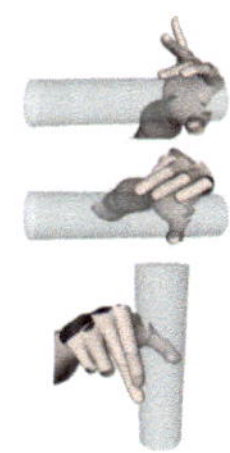	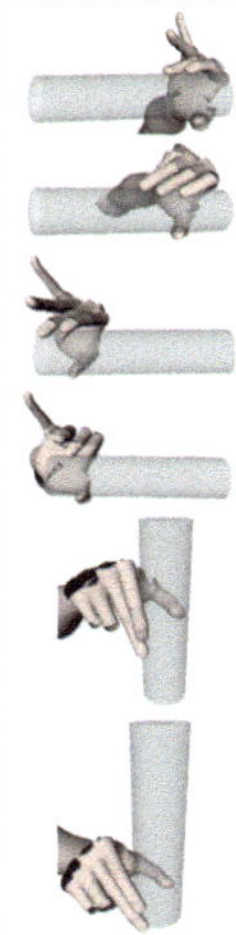
Min	0.0556 	0.0353 	0.8856 	0.8355 	0.2286 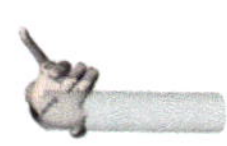	0.1680 
Max	0.5315 	0.2401 	0.9996 	0.9984 	0.9624 	0.9183 

rates the claw grasp better, although these rate values are not significantly different from the values of the cylindrical and lateral ones.

In conclusion, evaluating these quality measures for a different object's weight produced changes in their variability ranges. Q_{C3} increased its maximum from 0.24 to 0.53. The biomechanical indices also slightly increased their maximums approaching them more to 1 (Q_{E1_N} from 0.9984 to 0.9996 and Q_{E2_N} from 0.9183 to 0.9624).

7.3.2.2 Varying Sizes

The results of the quality measures for grasping cylinders with different diameters are presented in Fig. 7.8 and the minimum and maximum values in Table 7.3.

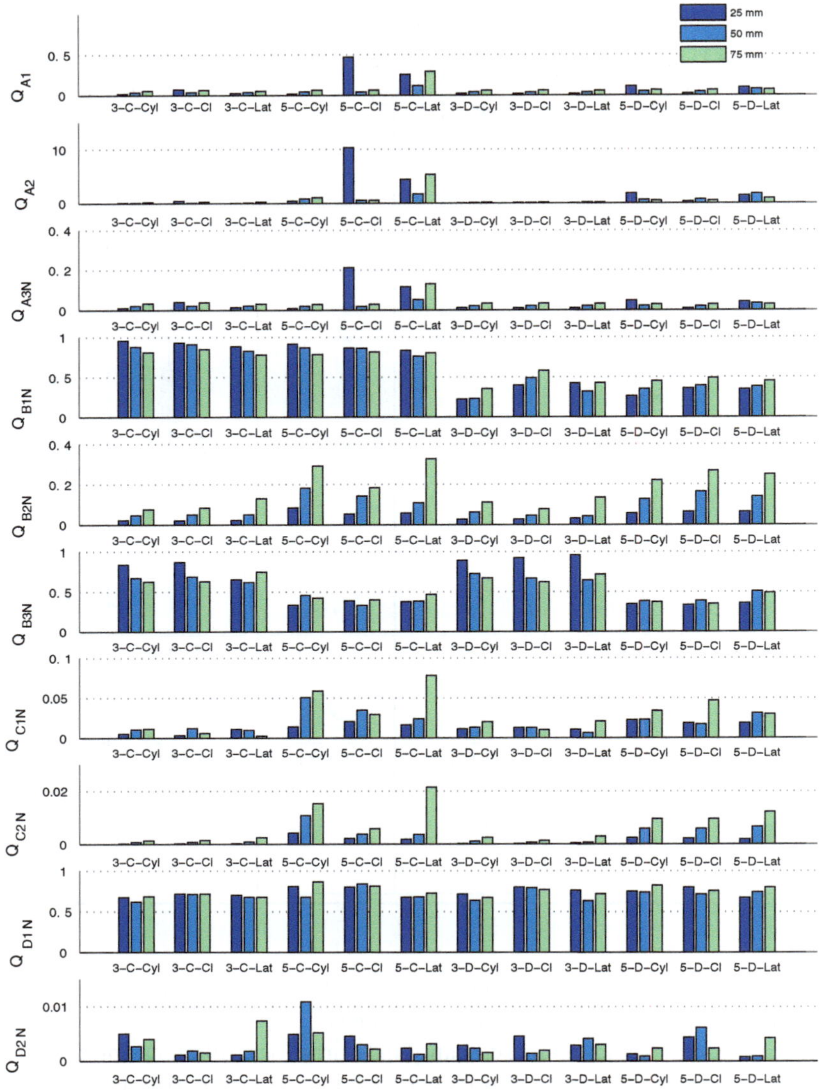

Fig. 7.8 Quality measures evaluated for 12 vertical grasps for cylinders of different diameter. Postures described by: Number of fingers (3, 5)—Position (*C* Centre, *D* Down)—Grasp type (*Cyl* Cylindrical, *Cl* Claw, *Lat* lateral)

Measures from Group A rated better claw and lateral 5-finger grasps in the centre, specially for the thin cylinder. Measure Q_{B1_N} showed better values for grasp in the centre than in the extremes, as expected. In the center, the thinner the cylinder the better, while in the extremes, grasps of thick cylinders were better rated. Q_{B2_N},

Table 7.3 Posture with the minimum and maximum grasp quality value for each measure when varying the object's size

	Q_{A1}	Q_{A2}	Q_{A3N}	Q_{B1N}	Q_{B2N}	Q_{B3N}	Q_{C1N}	Q_{C2N}	Q_{D1N}	Q_{D2N}
Min	0.0177 	0.0235 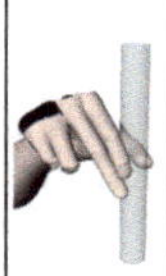	0.0084 	0.2219 	0.0214 	0.3333 	0.0025 	0.0002 	0.6172 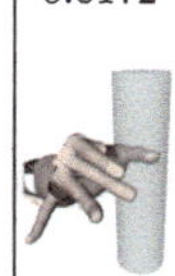	0.0008 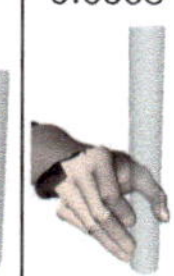
Max	0.4722 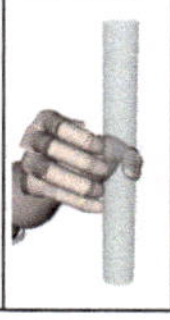	10.2706 	0.2111 	0.9578 	0.3273 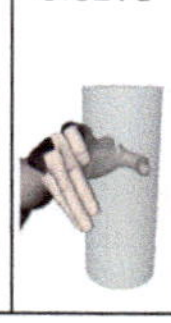	0.9597 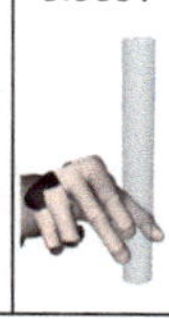	0.0779 	0.0214 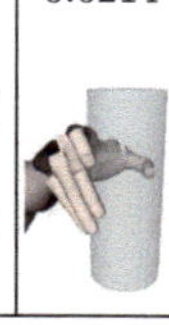	0.8655 	0.0109 

which measures the area of the grasp polygon, clearly increased with the cylinder thickness. In contrast, the regularity of the grasp polygon, measured by Q_{B3_N}, usually increased as the thickness of the cylinder decreased. Group C measures rated better 5 than 3-finger grasps. They also increased with the thickness of the cylinders which means that these grasps' ability to resist external wrenches is better. Measure Q_{D1_N} generally rated with high values (from 0.61 to 0.87) all grasps, giving better values to five-finger grasps that are closer to the relaxed posture. Finally, there is no clear pattern between the size and measure Q_{D2_N}. It rated worst the grasp performed with 5 fingers of the thinner cylinder which is a difficult grasp, and better the one performed with five fingers of the medium cylinder from the center which could also be expected.

In conclusion, evaluating these quality measures for different object's diameters produced changes in their variability ranges. In this case, not only their maximums values increased but also their minimums. Ranges from measures from Groups C and D remained practically unchanged. In contrast, all measures from Groups A and B suffered significant increments, being the greater the range from Q_{B3_N} which increased a 20 % of the total range.

7.3.2.3 Varying Fingers

The results of the quality measures for grasping cylinders with different fingers are presented in Fig. 7.9 and the minimum and maximum values are presented in Table 7.4.

Group A measures had a similar behaviour, rating better grasping the cylinder vertically from the bottom than from the top or horizontally. Grasps with four fingers excluding the index and with 5 fingers were preferred. The distance from the center

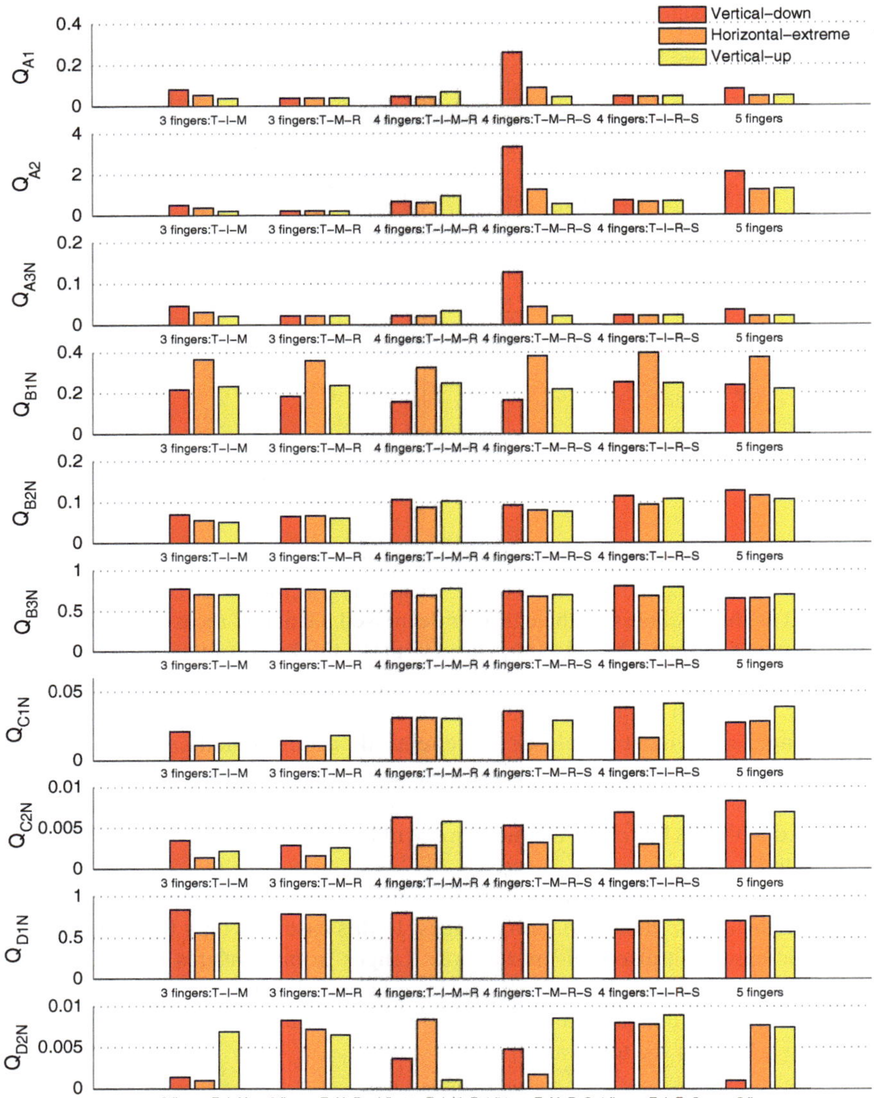

Fig. 7.9 Quality measures evaluated for 18 top grasps varying the number of fingers used to perform the grasp (*T* Thumb, *I* Index, *M* Middle, *R* Ring and *S* Small)

of mass, measured with Q_{B1_N}, increased for horizontal grasps and with the number of fingers. Measure Q_{B2_N} increased with the number of fingers but Q_{B3_N} remained similar for 3 and 4 finger grasps but less when using 5 fingers. Measures Q_{C1_N} and Q_{C2_N} increased with the number of fingers and were usually better for vertical

Table 7.4 Posture with the minimum and maximum grasp quality value for each measure when varying the fingers used to perform the grasp

	Q_{A1}	Q_{A2}	Q_{A3N}	Q_{B1N}	Q_{B2N}	Q_{B3N}	Q_{C1N}	Q_{C2N}	Q_{D1N}	Q_{D2N}
Min	0.0382 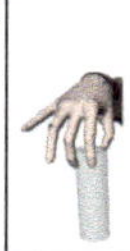	0.2082 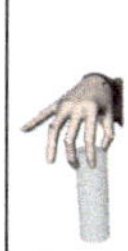	0.0212 	0.1580 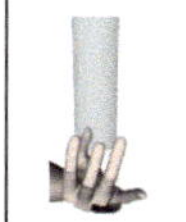	0.0515 	0.6494 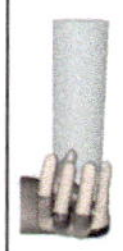	0.0107 	0.0014 	0.5639 	0.0010 
Max	0.2597 	3.3465 	0.1285 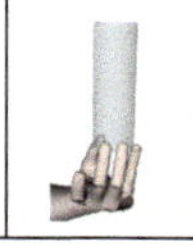	0.3965 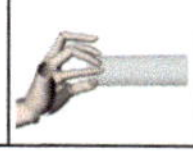	0.1274 	0.8062 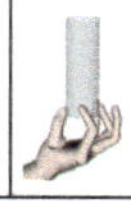	0.0413 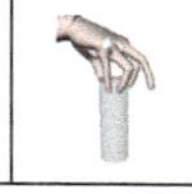	0.0083 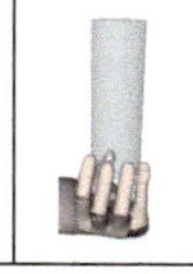	0.8434 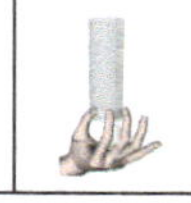	0.0089 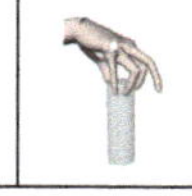

than horizontal grasps. Q_{D1_N} continued to give high values to all grasps (from 0.56 to 0.84) but there is no a clear pattern for the values when changing the number of fingers or the object orientation. Measure Q_{D2_N} showed significant variation in these experiments given that the manipulability of the grasps was highly affected when changing the fingers used to perform the grasp. It assigned better manipulability to grasps with four fingers excluding the middle in all orientations and worst to grasp the cylinder in vertical orientation from its bottom with 5 fingers. Three finger grasps with the thumb, middle and ring seemed to have better manipulability than using the index, and for 4-finger grasps it changed the values depending on the fingers used. In general, measures rated better vertical than horizontal grasps, from the bottom than from the top and with 4 fingers.

To conclude, in contrast with the previous variations studied, varying the number of fingers used in the grasp didn't significantly affect the original found ranges. The minimums of Q_{B1_N} and Q_{D1_N} were slightly reduced by 0.07 and 0.02 respectively. Likewise, the maximums of Q_{A3_N} and Q_{B3_N} were slightly increased by 0.01 and 0.05 respectively.

7.3.3 Discussion

Changing the number of fingers or the posture, weight or size of the object changed the ranges for each measure. The object weight specially increased the maximum of Q_{C3} but also slightly of Q_{E1_N} and Q_{E2_N}. The object's size varied the ranges of all measures specially the ones from Groups A and B. Lastly, the number of fingers only slightly decreased or increased the limits for few measures. Table 7.5 presents a summary of the minimum and maximum values of each measure over all the performed experiments.

Table 7.5 Summary of range limits found for each measure

	Q_{A1}	Q_{A2}	Q_{A3N}	Q_{B1N}	Q_{B2N}	Q_{B3N}	Q_{C1N}	Q_{C2N}	Q_{C3}	Q_{D1N}	Q_{D2N}	Q_{E1N}	Q_{E2N}
Min	0.0177	0.0140	0.0084	0.1580	0.0214	0.3333	0.0000	0.0002	0.0353	0.5639	0.0008	0.8355	0.1680
Max	0.4722	10.271	0.2111	0.9578	0.3273	0.9597	0.0779	0.0214	0.5315	0.8655	0.0109	0.9996	0.9624

Table 7.6 Proposed range for each measure

	Q_{A1}	Q_{A2}	Q_{A3N}	Q_{B1N}	Q_{B2N}	Q_{B3N}	Q_{C1N}	Q_{C2N}	Q_{C3}	Q_{D1N}	Q_{D2N}	Q_{E1N}	Q_{E2N}
Min	0.00	0.00	0.00	0.07	0.00	0.27	0.00	0.00	0.00	0.51	0.00	0.81	0.08
Max	0.52	11.30	0.30	1.00	0.36	1.00	0.09	0.03	0.59	0.90	0.02	1.00	1.00

Although all measures, except Q_{A1} and Q_{A2}, have been normalized, they seem to cover a small interval within the theoretical range of variation (0–1). Q_{B1N} and Q_{B3N} are varying well into the range. Q_{D1N} presents values with bias to 1, while the remaining normalized measures are biased to 0. It seems so that the mathematical limits chosen do not correspond to the achievable limits of variation of the measures in reality.

Therefore the minimum and maximum values found can help us determine the real range in which the quality measures vary, and better normalize them. It seems unlikely, even though a larger set of possible grasps was considered, that they cover the entire range. However, to account for possible variations provided by the consideration of other grasps, we propose to increase these ranges for their use in the normalization of the measures: the maximum and minimum values of each measure have been increased and decreased a 10 % of the range of the measure, respectively. Additionally numbers were rounded to the second decimal. The proposed ranges for each measure are shown in Table 7.6.

Figure 7.10 shows the comparison between the variability of the measures normalized with mathematical limits and with the proposed ranges for the 36 initial postures. It can be seen that the values are much better distributed between the range [0–1].

7.4 Sensitivity Analysis

It is important to determine which of the measures are robust, as the validity of the evaluations depends on their sensitivity to small variations in the input parameters used for their calculation. The adapted robotic indicators use the grasp posture as input parameter, and the simulator calculates the contact points and normal directions using a collision algorithm between the hand and the grasped object. The joint angles that define each grasping posture are commonly obtained experimentally registering the posture that human subjects use to grasp the given objects. This can be performed using different techniques as motion capture systems with high resolution optical cameras (e.g. Vicon system) or instrumented gloves (e.g. CyberGlove®, 5DT Data Glove). All techniques have uncertainty errors in their measurements, which are in the order of 3–6° [8–11]. Therefore, the quality measures calculated can be affected by these errors.

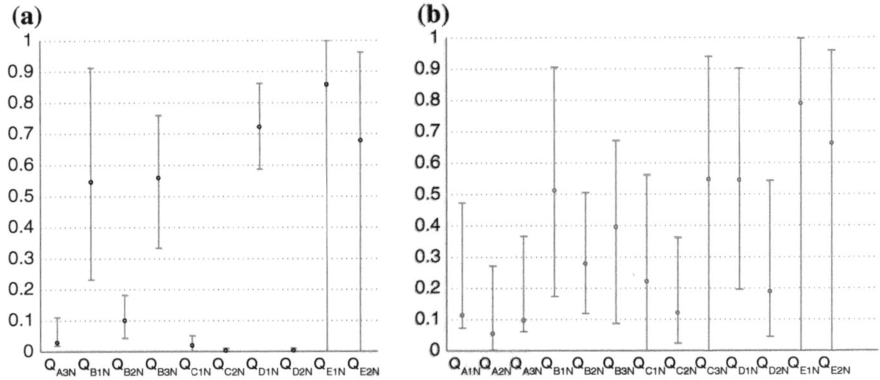

Fig. 7.10 Variability ranges over the 36 initial postures. **a** Normalized with mathematical ranges. **b** Normalized with proposed ranges

7.4.1 Material and Methods

In order to determine how much each measure was affected by small variations in the input grasp posture, a sensitivity analysis was performed. The experiments were selected choosing the 12 last grasp postures initially considered (Fig. 7.3) which corresponds to three and five-finger vertical grasps from the cylinder's centre and bottom using different grasp postures. Each of the initial postures was considered as the reference posture, and we introduced a variation in each one of their joints using a random sampling in an interval of ±5 % of its range of motion. We used 1,000 variations of the original posture for each cylinder. Only ten quality measures were used, excluding Q_{C3}, Q_{E1_N} and Q_{E2_N} which required more computational time as they need to run the muscular model. The evaluation of the measures was performed for each variation and its sensitivity was calculated analysing its deviation from the value obtained for the reference posture. A Global Sensitivity Index (GSI) for each measure was obtained as the root-mean-square deviation (RMSD) with respect to measure calculated for the reference posture:

$$GSI = \sqrt{\left(\frac{1}{n-1}\sum_{i=1}^{n}(x_i - x_0)^2\right)} \tag{7.1}$$

where n is the number of variations, x_i the value of the measure calculated for each variation of grasp posture and x_o the value of the measure of the reference posture.

Each *GSI* has the same units as each measure. In order to normalize *GSI* and give the result in percentage, the range of each measure has been used:

$$GSI_N = \frac{GSI}{max - min} \tag{7.2}$$

where max and min were selected using two approaches: first the value of 1 and 0 for the measures that are normalized using their mathematical limits, and second, using the ranges obtained from the variability analysis.

7.4.2 Results

The values of the global sensitivity index obtained for the 10 measures are shown in Fig. 7.11. However, the values do not have any meaning if they are not compared with the range of variation of each measure.

For the measures which are normalized with their mathematical limits (all except Q_{A1} and Q_{A2}), we compared the GSI with the range between 0 and 1 and showed the results in percentage for each measure (Fig. 7.12). The graph shows that none of the indicators varied more than 10 % for all measures, with Q_{B3_N} and Q_{D1_N} being the most sensitives. But as the variation was made in 10 % of the range, we could conclude that these measures would be robust to uncertainties in the initial grasp posture.

However, as we already verified that the mathematical limits that we used to normalize the measures between 0 and 1 were not realistic, a more meaningful evaluation of the sensitivity would be obtained comparing the GSI with the proposed range found after the variability study. The results for the 10 measures are shown in Fig. 7.13.

The graph shows that all measures increased the sensitivity in comparison with the GSI with the adjusted range. Their means are below 12 % and their maximums are all below 25 % of their range of variation. Measures Q_{D1_N} and Q_{D2_N} shows their means close to 10 %, but their maximums were the highest ones, which should be considered

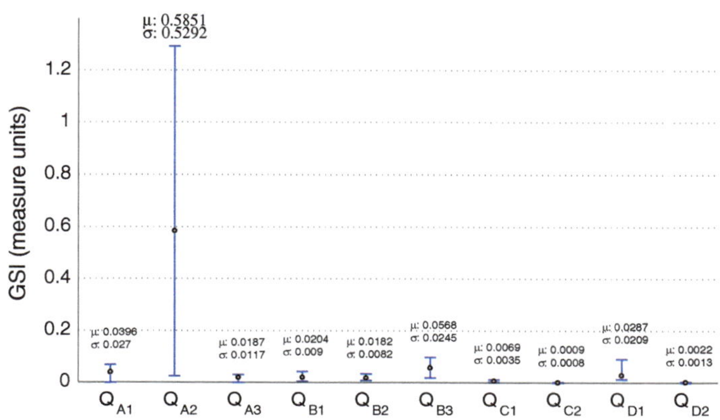

Fig. 7.11 GSI shown in each measure units. The mean (μ) and the standard deviation (σ) over all selected postures are shown for each measure

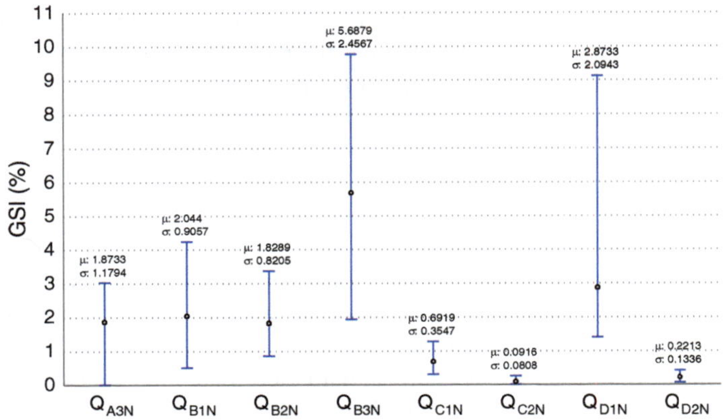

Fig. 7.12 GSI shown in percentage over the range between 0 and 1. The mean (μ) and the standard deviation (σ) over all selected postures are shown for each measure

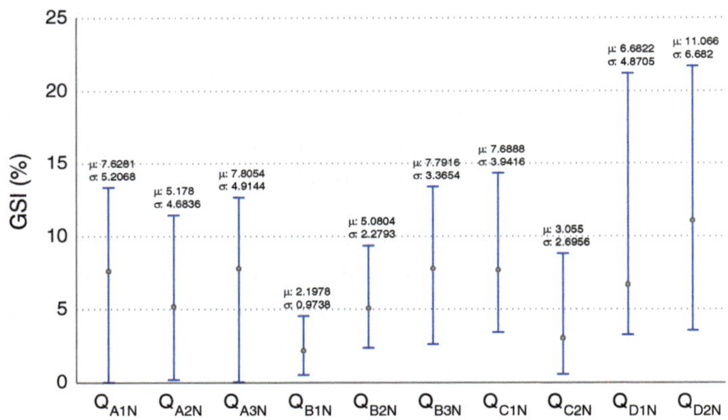

Fig. 7.13 GSI shown in percentage over the range proposed in the variability analysis. The mean (μ) and the standard deviation (σ) over all selected postures are shown for each measure

when using these measures. However, it should be noted that the posture of the hand was the parameter varied in the experiments, therefore these two measures should be the most affected as they consider the configuration of the manipulator. Measure Q_{B1_N} is the most robust followed by Q_{C2_N}. Nevertheless, a 12 % of variation shows that none of the indicators are too sensitive to changes in the input grasp posture.

Table 7.7 Results of the statistical correlation between different quality measures (Light Obj. = 180 g)

	Q_{A1}	Q_{A2}	Q_{A3_N}	Q_{B1_N}	Q_{B2_N}	Q_{B3_N}	Q_{C1_N}	Q_{C2_N}	Q_{C3}	Q_{D1_N}	Q_{D2_N}	Q_{E1_N}	Q_{E2_N}
Q_{A1}	1.0												
Q_{A2}	0.8	1.0											
Q_{A3_N}	1.0	0.6	1.0										
Q_{B1_N}	0.3	0.3	0.3	1.0									
Q_{B2_N}	0.2	0.6	0.1	0.2	1.0								
Q_{B3_N}	−0.2	−0.4	−0.1	−0.1	−0.8	1.0							
Q_{C1_N}	0.1	0.3	0.0	0.2	0.7	−0.8	1.0						
Q_{C2_N}	0.2	0.2	0.0	0.1	0.9	−0.7	0.7	1.0					
Q_{C3}	0.1	0.2	0.1	0.4	0.3	−0.2	0.2	0.2	1.0				
Q_{D1_N}	0.0	0.0	0.0	0.0	0.3	−0.5	0.5	0.1	0.0	1.0			
Q_{D2_N}	−0.3	−0.1	−0.4	−0.1	0.4	−0.2	0.3	0.5	0.0	−0.1	1.0		
Q_{E1_N}	−0.4	−0.2	−0.4	−0.1	0.1	−0.1	0.1	0.0	0.6	0.1	0.2	1.0	
Q_{E2_N}	−0.3	0.0	−0.4	−0.1	0.3	−0.3	0.3	0.2	0.6	0.1	0.3	0.9	1.0

7.5 Independent Grasp Aspects

In order to analyse the relationships between the quality measures, the Pearson correlation coefficient was calculated for each combination of measures. These correlations enable us to determine sets of measures that evaluate similar aspects of the grasp. The identification of groups of correlated measures can help us to reduce the number of indices that need to be calculated in order to assess the quality of a given grasp. For each of the independent sets of measures identified, a physical aspect can be associated which describes the aspect being measured. Moreover, one measure from each of these groups can be selected, given that it would be enough to assess these aspects.

The 36 initial postures were considered for this analysis, for the light and heavy cylinder. The results of the correlations for each combination of measures are shown in Table 7.7 for the lighter cylinder (180 g, 50 cm in diameter). Three different ranges of correlations have been considered and marked in the table. Perfect correlations (1.0) have been marked in dark grey, high correlations ($\geq$0.7) have been marked in medium grey and moderate correlations ($\geq$0.5) have been marked in light grey.

Changing the object weight only modifies measures Q_{C3} and Q_{E1}; therefore, the results of the correlations for these measures with the heavier cylinder (460 g) are shown in Table 7.8.

Measures from Group *A* showed a high correlation with each other, which leads us to conclude that calculating only one of them provides us with the same evaluation of a given grasp. From these three, the measure Q_{A3_N} is preferable given that it is already normalized.

The measure Q_{B1_N} showed no correlation with any other measure which means that measuring stability by avoiding inertial forces when reducing the distance be-

Table 7.8 Results of the statistical correlation between different quality measures (Heavy Obj. = 460 g)

	Q_{A1}	Q_{A2}	Q_{A3N}	Q_{B1N}	Q_{B2N}	Q_{B3N}	Q_{C1N}	Q_{C2N}	Q_{C3}	Q_{D1N}	Q_{D2N}	Q_{E1_N}	Q_{E2_N}
Q_{C3}	0.5	0.7	0.5	0.5	0.6	−0.5	0.4	0.5	1.0	0.0	0.0	0.6	0.8
Q_{E1N}	0.2	0.3	0.1	0.2	0.4	−0.3	0.3	0.3	0.6	0.2	0.1	1.0	0.8
Q_{E1N}	0.2	0.5	0.1	0.2	0.7	−0.6	0.5	0.6	0.8	0.2	0.2	0.8	1.0

tween the centroid of the contact polygon and the object's centre of mass is an independent aspect of the grasp not evaluated by any of the remaining indices. In contrast, measures Q_{B2_N} and Q_{B3_N} were highly correlated with each other and with measures Q_{C1_N} and Q_{C2_N} which also measure stability in terms of the force that a grasp can resist. Additionally Q_{B2_N} was weakly correlated with the stability measure Q_{A2_N}, showing that a grasp is more stable when the area of the grasp polygon is bigger; and Q_{B3_N} was inversely correlated with Q_{D1_N}, showing that in order to achieve a more perfect shape for the grasp polygon, the hand needs to move the fingers to a configuration in which the joints are closer to their limits, which can lead to a less manipulable and uncomfortable grasp.

The measures from Group C, Q_{C1_N} and Q_{C2_N}, were correlated with each other and with measures Q_{B2_N} and Q_{B3_N}, as was mentioned earlier, which shows that the grasp stability is related to the shape of the grasp polygon. Additionally, they presented weak correlations with Q_{A2} which is also a stability indicator.

The measures from Group D seem to measure different aspects of the grasp because they were only weakly correlated with other measures. Q_{D1_N} was correlated with Q_{C1_N} and inversely with Q_{B3_N}; and Q_{D2_N} was not correlated with any measure.

Finally, the biomechanical indices Q_{E1_N} and Q_{E2_N} were highly correlated with each other and with Q_{C3}—the only additional measure which needs the biomechanical model. These measures increased their correlations when grasping a heavier object: Q_{C3} with measures from Group A and B; and Q_{E2_N} with Q_{B2_N}, Q_{B3_N} and measures from Group C. Q_{E2_N} however maintained similar high correlations for both weights.

These correlations show that there are measures that evaluate similar aspects of the grasp. Therefore, several groups of correlated measures can be identified in order to reduce the number of indices that need to be calculated in order to assess the quality of a given grasp:

- Group 1: Q_{A1}, Q_{A2}, Q_{A3_N}
- Group 2: Q_{B2_N}, Q_{B3_N}, Q_{C1_N} and Q_{C2_N}
- Group 3: Q_{B1_N}
- Group 4: Q_{D1_N}
- Group 5: Q_{D2_N}
- Group 6: Q_{E1_N}, Q_{E2_N} and Q_{C3}.

For each of the six independent sets of measures that have been identified above as evaluating different aspects of the grasp, a physical interpretation can be associated which describes the aspect being measured. Moreover, one measure from each of these groups is selected as representative to evaluate this aspect.

7.5.1 Restriction of the Grip

The first group is composed of stability indicators, which give an idea of how restricted the grip is. Q_{A3_N} can be chosen to calculate this aspect given that it is already normalized and has been found to be robust enough.

7.5.2 Ability to Resist Forces

The second group is composed of measures from Group B, Q_{B2_N}, Q_{B3_N} and two measures from Group C, Q_{C1_N} and Q_{C2_N} which give an idea of the grasp's ability to resist external wrenches. From this group, Q_{C1_N} can be chosen given that it is the most common measure used in robotics and does not present problems of robustness.

7.5.3 Dynamic Effects

The third group is composed only by the measure Q_{B1_N} which gives an idea of the grasp ability to counteract dynamic effects. This measure gives priority to grasps closer to the centre of mass and penalizes the ones closer to the object's extremities which are more likely to be affected by inertial forces. The experiments defined in this work only consider static grasps which are not affected by dynamic effects, therefore this measure does not have a high impact. Other experiments considering transport or manipulation tasks could benefit more from the information provided by this measure.

7.5.4 Comfort

The fourth group is composed of only the measure Q_{D1_N}. When the hand joints are at their limits, their soft tissues are subjected to large deformation, generating discomfort. Therefore, this index measures how comfortable a grasp is, assuring that the finger joints are far from their limits. This aspect also gives an idea of manipulability of the grasp in a way, because in order to make changes in the hand posture, the joints should have an available range of movement which does not happen when they are at their limits.

7.5.5 *Manipulability*

The fifth group is composed of only the measure Q_{D2_N} which gives an idea of the manipulation ability measuring the capability of the hand to move an object in any direction.

7.5.6 *Muscular Fatigue*

The sixth group is composed of Q_{C3} and the biomechanical indices Q_{E1_N} and Q_{E2_N}. From this group, Q_{E1_N} is chosen given that it is easier to calculate than Q_{E2_N} and its low correlation with the rest of the measures, showing that it is measuring a completely new aspect of the human grasp.

This set of independent aspects can then be used to evaluate different grasps and enable their comparison. Radar plots have been used in Fig. 7.14 to show an example of how to represent the six independent aspects for different sets of grasps.

7.6 Validation of Grasp Quality Measures

The following step consists on the validation of the results obtained using the selected quality measures. In order to do that, we use the human innate ability that allows them to evaluate different grasps, compare them and give an overall assessment of their quality.

7.6.1 *Human Assessment*

A set of experiments were designed to ask subjects their assessment of different grasps and use them as the target values to evaluate the results of the proposed grasp quality measures.

The light (180 g) and heavy (460 g) cylinders (used in the previous section) were grasped using the 36 different postures shown in Fig. 7.3 to assess the validity of the global quality measure. Twenty four female subjects were selected with ages ranging between 18 and 60 years and without hand pathologies. Their hand anthropometric dimensions were chosen to be within the 25–75 percentiles among a random sample of 60 women, laying in the range from 170 to 179 mm in HL (length) and 73 to 78 mm in HB (breadth). They were asked to grasp the cylinder imitating a specific grasp shown in a picture in which we varied the grasp type (cylindrical, claw and lateral), orientation (horizontal and vertical) and position (centre, up and down for vertical orientation; centre, right and left for horizontal orientation).

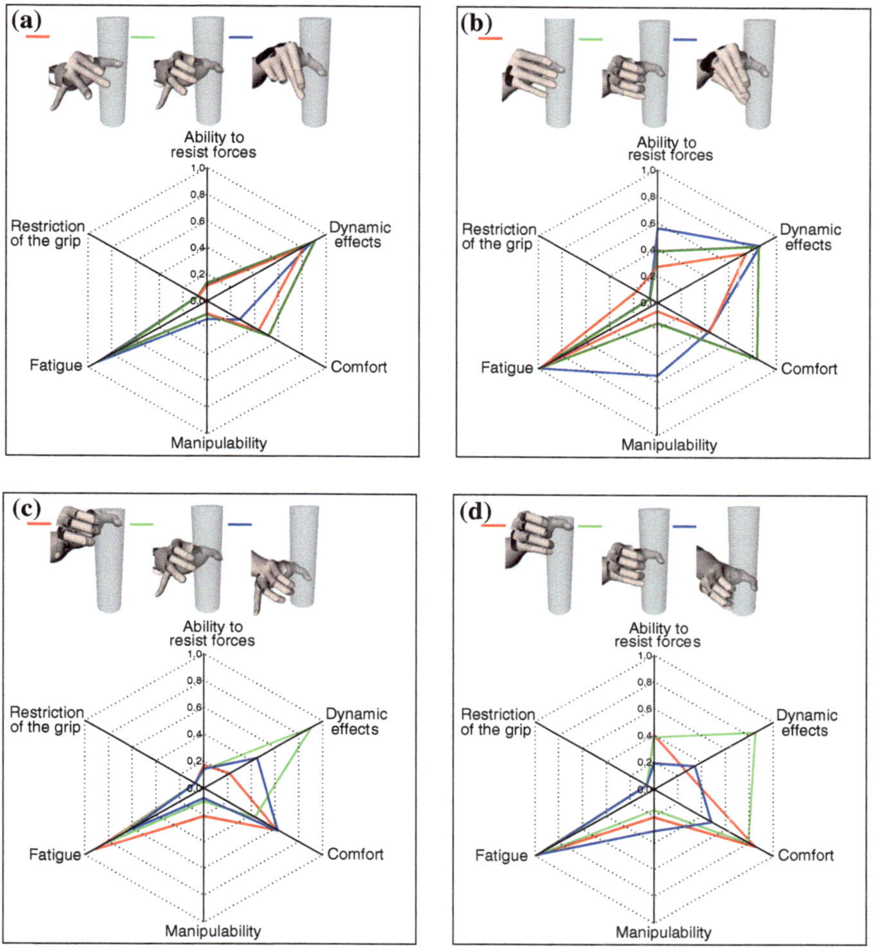

Fig. 7.14 Comparison of independent grasp aspects. **a** Comparison between postures (cylindrical, claw, lateral) with 3 fingers. **b** Comparison between postures (cylindrical, claw, lateral) with 5 fingers. **c** Comparison between vertical grasps (up, centre, down) with 3 fingers. **d** Comparison between vertical grasps (up, centre, down) with 5 fingers

Since it is not feasible to ask a subject to perform 36 different grasps and subsequently rate them all, we grouped them into two lists of six groups, each group containing six combinations (Table 7.9).

These groups were distributed randomly between the subjects. Finally, each subject evaluated six combinations which means that each combination is evaluated by four different subjects. For each experiment, the subject was asked to perform six grasps for a transport task and, at the end, assess their quality by ordering them from

Table 7.9 Combination of experiments

Constants	Variables
Horizontal, Left	# fingers, Grasp type
Horizontal, Centre	# fingers, Grasp type
Horizontal, Right	# fingers, Grasp type
Vertical, Up	# fingers, Grasp type
Vertical, Centre	# fingers, Grasp type
Vertical, Down	# fingers, Grasp type
3 fingers, Cylindrical	Position, Orientation
3 fingers, Claw	Position, Orientation
3 fingers, Lateral	Position, Orientation
5 fingers, Cylindrical	Position, Orientation
5 fingers, Claw	Position, Orientation
5 fingers, Lateral	Position, Orientation

highest to lowest in terms of comfort or ease of grasp. This procedure was performed for both cylinders.

A statistical analysis, using a Tukey's honestly significant difference (HSD) test, was performed with the human assessment for each cylinder and a human global quality measure (GQM) was assigned ranging between 1 and 6, corresponding to the best and worst grasp respectively. As the experiments were selected using combinations of grasps with the same cylinder, there was no information to create a global ranking of both weights. However, 3 subjects were asked to compare the worst assessed grasp for the light cylinder with the best grasp for the heavy one. In all cases, subjects preferred the worst of the light cylinder grasp than the best of the heavy cylinder grasp. Therefore, the human grasp quality measure of the heavy cylinder has been increased by 6 to rate them worse than the light's one. We normalized the human assessment to obtain a global quality measure in the range between 0 and 1, and modified it to have 1 as its best value: $Human\, GQM_N = 1 - (GQM - 1)/(12 - 1)$.

The results of ranking the 36 grasps for each cylinder is presented in Table 7.10 and are used in the following section to validate the different independent grasp quality measures.

7.6.2 Comparison with Quality Measures Results

The subjects' assessment of the different grasps were compared with the results of the proposed grasp quality measures. Figure 7.15 shows the results for each grasp independent aspect evaluated for the 72 postures (36 per cylinder). The postures were ordered by the human assessment and their expected value is shown in black. The results obtained using each independent aspect is depicted in red and its Pearson's correlation coefficient with the subjects' assessment is presented.

Table 7.10 Ranking of grasp quality assessed by human subjects (best to worst)

Light cylinder						Heavy cylinder					
F[a]	Grasp type	Orientation	Position	GQM[b]	GQM^c_N	F[a]	Grasp type	Orientation	Position	GQM[b]	GQM^c_N
5	Lateral	Vertical	Down	1.75	0.93	3	Cyl.	Horizontal	Centre	1.75	0.39
3	Cyl.	Horizontal	Centre	1.75	0.93	3	Cyl.	Vertical	Centre	2.00	0.36
3	Cyl.	Vertical	Up	2.00	0.91	5	Cyl.	Vertical	Centre	2.00	0.36
5	Cyl.	Horizontal	Right	2.25	0.89	5	Cyl.	Vertical	Up	2.00	0.36
3	Cyl.	Vertical	Centre	2.50	0.86	5	Cyl.	Vertical	Down	2.25	0.34
3	Claw	Vertical	Centre	2.50	0.86	5	Cyl.	Horizontal	Centre	2.25	0.34
5	Claw	Vertical	Centre	2.50	0.86	5	Claw	Vertical	Centre	2.40	0.33
5	Cyl.	Vertical	Up	2.50	0.86	5	Lateral	Vertical	Down	2.40	0.33
3	Lateral	Vertical	Down	2.50	0.86	5	Lateral	Vertical	Centre	2.60	0.31
5	Cyl.	Vertical	Down	2.50	0.86	3	Cyl.	Vertical	Up	2.75	0.30
3	Lateral	Vertical	Centre	2.75	0.84	5	Lateral	Vertical	Up	2.80	0.29
5	Cyl.	Horizontal	Centre	3.00	0.82	3	Lateral	Horizontal	Centre	3.00	0.27
5	Claw	Horizontal	Left	3.00	0.82	3	Lateral	Horizontal	Left	3.00	0.27
5	Lateral	Vertical	Centre	3.25	0.80	5	Claw	Vertical	Up	3.20	0.25
3	Claw	Vertical	Up	3.25	0.80	3	Claw	Horizontal	Centre	3.25	0.25
3	Cyl.	Vertical	Down	3.25	0.80	5	Cyl.	Horizontal	Left	3.25	0.25
5	Cyl.	Vertical	Centre	3.50	0.77	5	Claw	Horizontal	Centre	3.40	0.24
5	Claw	Vertical	Up	3.50	0.77	3	Cyl.	Vertical	Down	3.50	0.23
5	Claw	Horizontal	Centre	3.50	0.77	3	Lateral	Vertical	Down	3.50	0.23
3	Lateral	Vertical	Up	3.75	0.75	3	Claw	Horizontal	Right	3.50	0.23
3	Cyl.	Horizontal	Right	3.75	0.75	5	Lateral	Horizontal	Centre	3.60	0.22
3	Cyl.	Horizontal	Left	3.75	0.75	3	Claw	Vertical	Centre	3.75	0.20

(continued)

Table 7.10 (continued)

Light cylinder						Heavy cylinder					
F[a]	Grasp type	Orientation	Position	GQM[b]	GQM_N^c	F[a]	Grasp type	Orientation	Position	GQM[b]	GQM_N^c
5	Lateral	Vertical	Up	4.00	0.73	5	Cyl.	Horizontal	Right	3.75	0.20
5	Lateral	Horizontal	Centre	4.00	0.73	3	Cyl.	Horizontal	Left	3.75	0.20
5	Lateral	Horizontal	Right	4.00	0.73	3	Claw	Vertical	Up	4.00	0.18
5	Cyl.	Horizontal	Left	4.00	0.73	3	Lateral	Vertical	Up	4.00	0.18
5	Lateral	Horizontal	Left	4.00	0.73	5	Claw	Horizontal	Right	4.00	0.18
3	Claw	Horizontal	Centre	4.25	0.70	5	Lateral	Horizontal	Left	4.20	0.16
3	Lateral	Horizontal	Centre	4.25	0.70	3	Lateral	Vertical	Centre	4.25	0.16
3	Claw	Horizontal	Right	4.50	0.68	5	Claw	Vertical	Down	4.40	0.15
5	Claw	Horizontal	Right	4.50	0.68	5	Lateral	Horizontal	Right	4.60	0.13
5	Claw	Vertical	Down	4.75	0.66	3	Cyl.	Horizontal	Right	5.00	0.09
3	Lateral	Horizontal	Left	5.00	0.64	3	Lateral	Horizontal	Right	5.00	0.09
3	Claw	Vertical	Down	5.25	0.61	3	Claw	Horizontal	Left	5.25	0.07
3	Lateral	Horizontal	Right	5.25	0.61	5	Claw	Horizontal	Left	5.40	0.05
3	Claw	Horizontal	Left	5.25	0.61	3	Claw	Vertical	Down	6.00	0.00

[a] Number of fingers used to perform the grasp
[b] GQM (Grasp quality measure) evaluated in the range between 1 (best) and 6
[c] GQM_N (Normalized Grasp quality measure) evaluated in the range between 0 (best) and 1

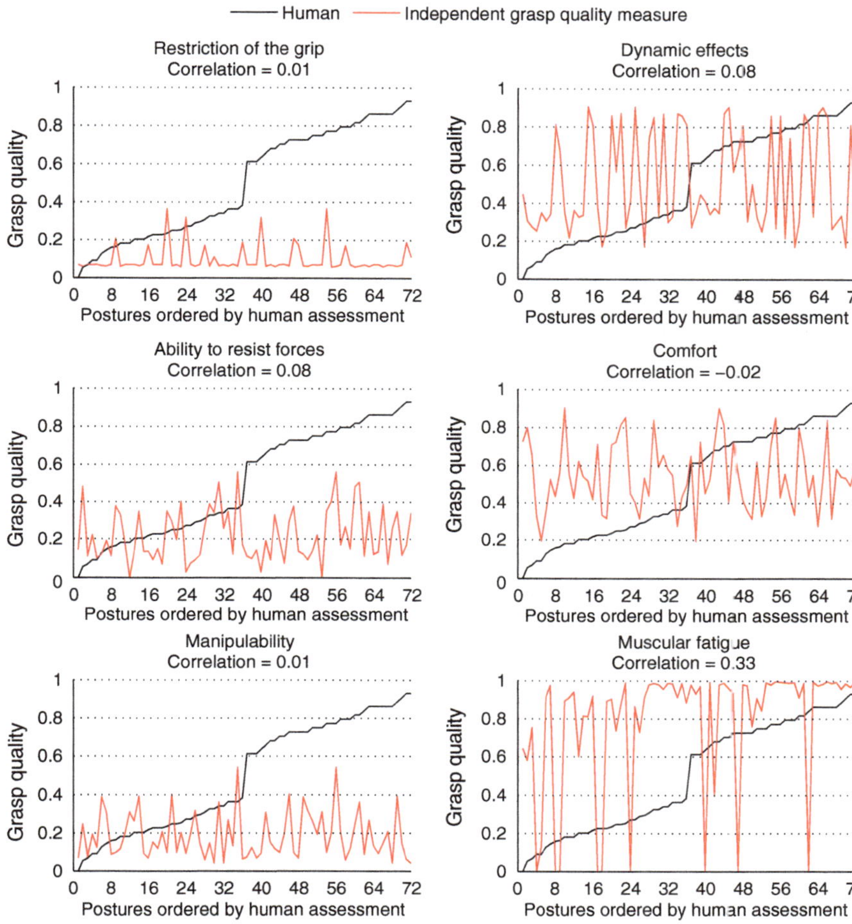

Fig. 7.15 Results of the grasp quality evaluation given by each independent aspect for the 72 postures ordered by the human assessment. The correlation coefficient with the human assessment is shown for each independent aspect

The correlation coefficient of the different independent aspects gave very low values (less than 0.5) showing the none of the aspects is able to, independently, predict the ranking assessed by the humans. This variety of the ranking results corroborates that the indices measure different aspects of the grasp, confirming the importance of combining them to create an overall quality index.

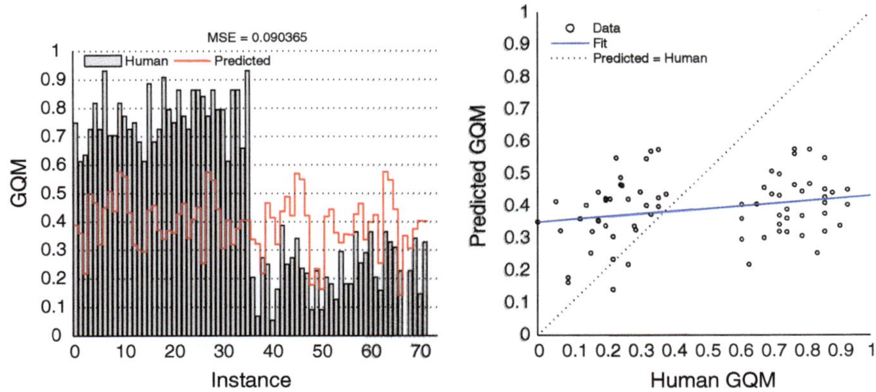

Fig. 7.16 Performance of merging the independent quality measures using their mean

7.7 Global Grasp Quality Measure

Different ways to perform the merging of the independent quality measures have been evaluated. We used the mean square error (MSE) of each proposed grasp quality measure compared with the human assessment as a measure of its performance. At the end, we selected the proposed grasp quality measure with the lowest MSE.

A first approximation was obtained by using the mean of the normalised qualities calculated by the different criteria, being the best grip the one with the largest score. The results of this average quality index are presented in Fig. 7.16 for the light and heavy cylinders.

The proposed GQM does not match the human assessment observed in the experiment. It overestimates the values for the light cylinder and underestimates them for the heavy one. Given that the only measure that considers the weight is the Fatigue index, it should influence the output more than the other measures. In order to change the influence of each criterion on the global quality value, it is necessary to use a more complex implementation such as a weighted sum. This approach has been used in previous studies [3, 4], however the difficulty lies in finding the appropriate weights for each measure. We propose to use artificial neural networks in which the independent quality measures are used as input and through a learning process the weights of the interconnections are updated to approximate the output to the human global quality measure.

Different one-layer feed-forward neural networks were trained with the Levenberg-Marquardt backpropagation algorithm. In this networks, the first layer has a connection from the network input and each subsequent layer has a connection from the previous layer. The final layer produces the network's output. They were implemented using the MATLAB Neural Network Toolbox and their general scheme is depicted in Fig. 7.17.

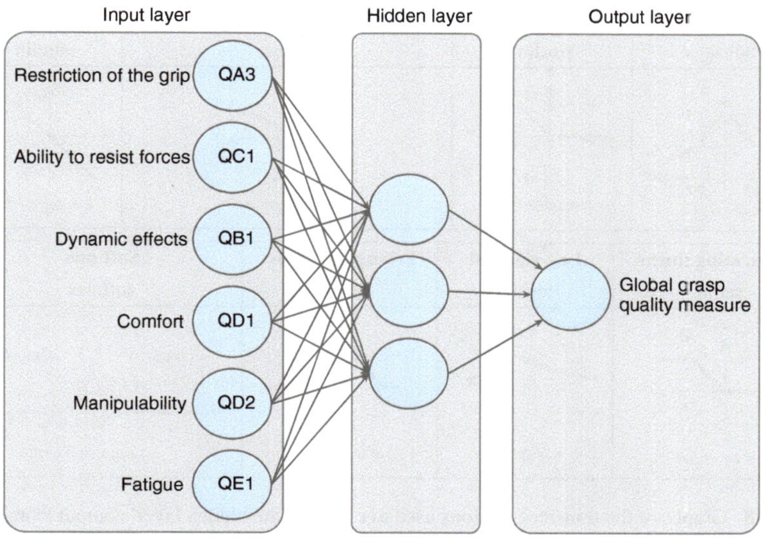

Fig. 7.17 Scheme of the neural network for the global grasp quality measure estimation

Combination of different parameters were tried to find the neural network that assigns appropriate weights to each independent aspect to best predict the human assessment:

- **Percentage of Data**: Randomly divide up the 72 samples
  - **Training**: 80 % (58 samples)
    These are input to the network during training and the network is adjusted according to its error.
  - **Validation**: 10 % (7 samples)
    These are used to measure network generalization, and to halt training when generalization stops improving.
  - **Testing**: 10 % (7 samples)
    These have no effect on training and so provide an independent measure of the network performance during and after training.

- **Transfer functions in the hidden layer**: They calculate the hidden layer's output from their net input. We used the eight functions shown in Fig. 7.18.
- **Transfer functions in the output layer**: They calculate the output layer's result from the hidden layer. We used the *Linear* function (purelin).
- **Number of neurons in the hidden layer**: We start with 10, which is the default number, and increase it by one until 30.
- **Number of training trials**: As training multiple times with the same parameters generate different results due to different initial conditions and sampling, we

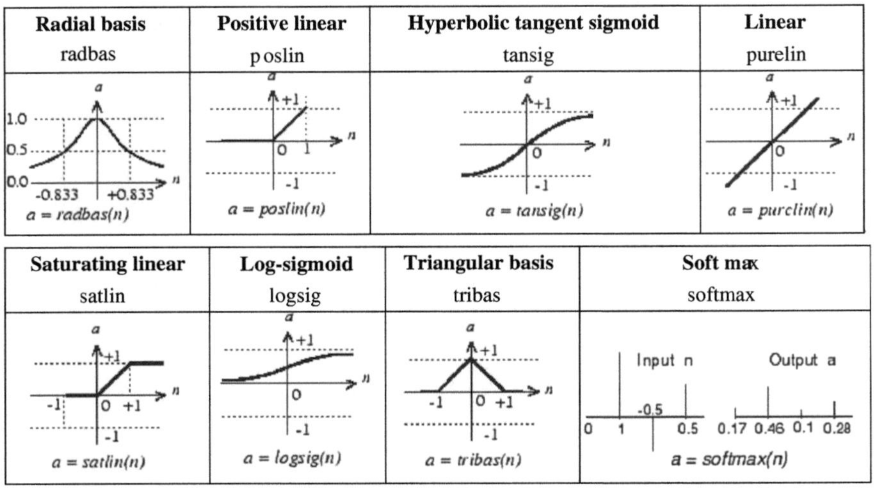

Fig. 7.18 Graphs of the transfer functions used to calculate the hidden layer's output from the net input. They were reproduced from the MATLAB Neural Network Toolbox documentation (http://www.mathworks.es/es/help/nnet/)

trained 50 times each combination of parameters and save the average performance error.

In order to measure the network's performance, we used the mean of squared errors (*MSE*). It is calculated as the average squared difference between outputs and targets. Lower values indicate better performance and zero means no error. Results of the average performance over 50 trials for each transfer function in the input layer and varying the number of neurons in the hidden layer are shown in Fig. 7.19.

A summary of the number of neurons which produced the best average performance for each transfer function is presented in Table 7.11. The network found with the best performance corresponds to a network with one hidden layer consisting of 29 neurons, using a *hyperbolic tangent sigmoid* transfer function which produced a mean squared error of 0.0284.

Using the saved neural network with the combination of parameters described, we have a way to evaluate the overall quality measure from the independent grasp aspects which matches the human assessment. Figure 7.20 shows the comparison of the net with the human assessment. Although there are still some outliers, it shows a better approximation to the human evaluation than the mean of the quality measures, decreasing the mean squared error from 0.090 to 0.028.

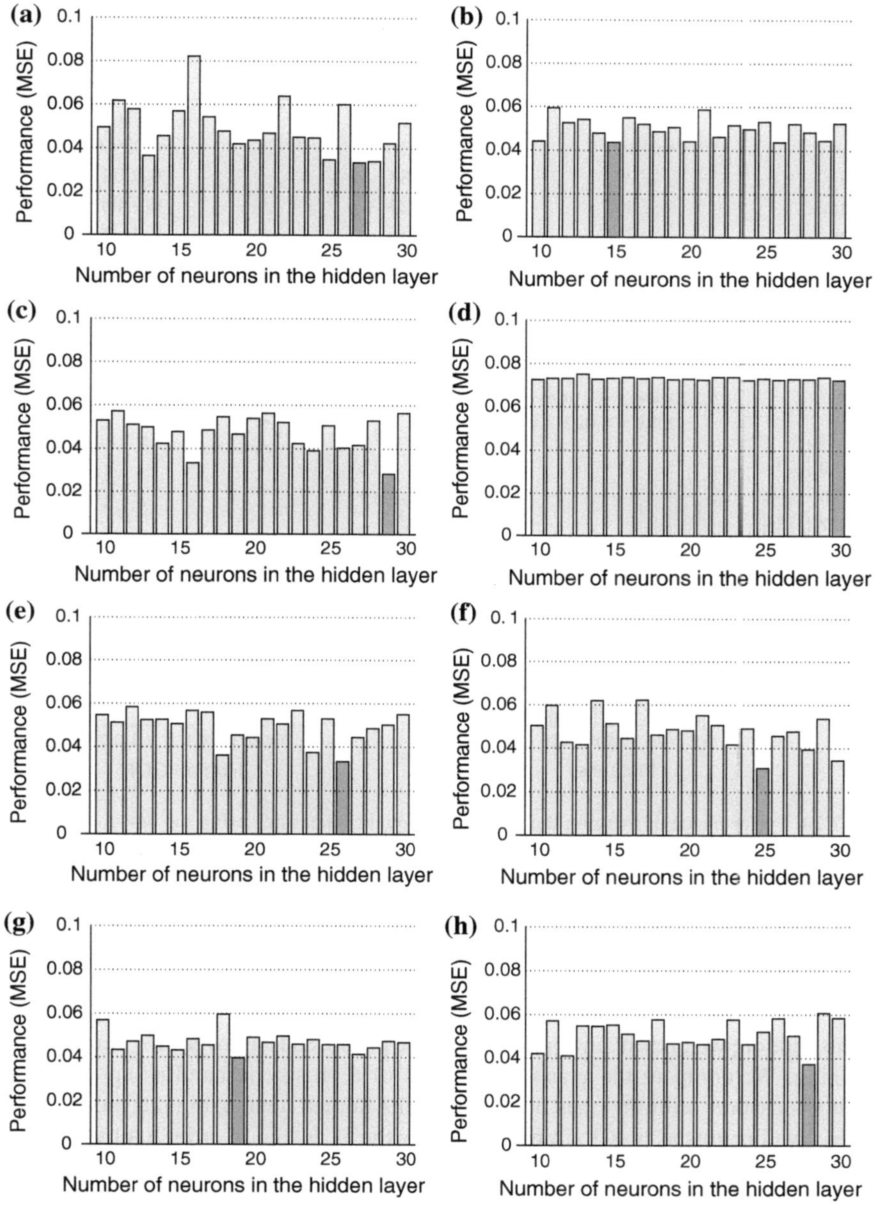

Fig. 7.19 Results of the average performance over 50 trials for each transfer function in the input layer and varying the number of neurons in the hidden layer. **a** Radbas, **b** poslin, **c** tansig, **d** purelin, **e** satlin, **f** logsig, **g** tribas, **h** softmax

Table 7.11 Number of neurons in the hidden layer which produced the best average performance for each transfer function

Transfer function	# Neurons	Average performance	Best performance
Radbas	27	0.0335	0.0202
Poslin	15	0.0436	0.0399
Tansig	29	0.0284	0.0231
Purelin	30	0.0724	0.0724
Satlin	26	0.0333	0.0228
Logsig	25	0.0308	0.0239
Tribas	19	0.0397	0.0293
Softmax	28	0.0373	0.0314

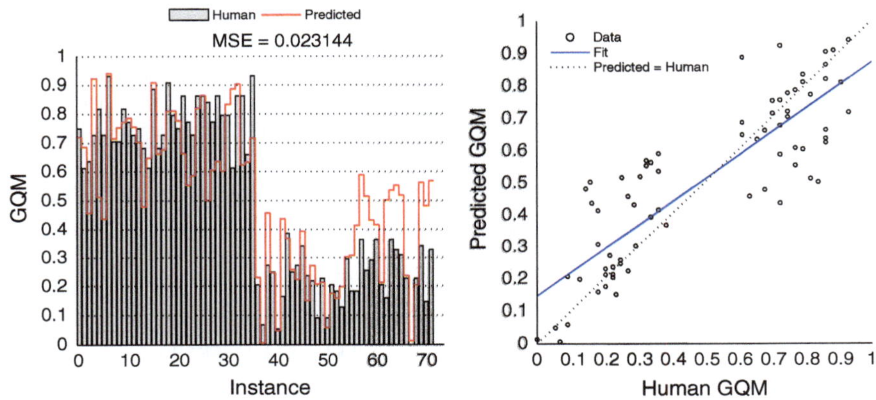

Fig. 7.20 Performance of merging the independent quality measures using the best found neural network

7.8 Comparative Evaluation of Prosthetic and Natural Human Hand Grasping

In this section, we present an application of the proposed framework for human hand grasping. We compare the grasp capabilities of a prosthetic hand using the grasp quality measures with the ones obtained for the human hand using our model. This could give insights as how to improve the prosthesis design in order to obtain better quality scores.

Prosthetics hands have evolved and improved over the years, helping people gaining manipulation capabilities. However, the current state of the art offers prosthesis that generally have a single degree of freedom (simultaneously opening and closing of joints), and are controlled using signals from only two muscles, captured by surface electrodes to perform the opening and closing movements. Although these devices are extremely robust and require little effort to learn its operation by the user, they have large deficiencies in their appearance and their functionalities are very limited. As a result, most available commercial devices do not meet the user expectations, thus these prostheses are often rejected by the amputee.

The most widely used and successful control method for the prosthesis is based on the myoelectric signal processing [12]. It consists in controlling the different degrees of freedom of the prosthesis relating the myoelectrical impulses generated by a user's muscle contractions, using a suitable control algorithm. There are several practical difficulties in reliably use more than two EMG channels to simultaneously control the prosthesis [13]. Thus, even the multi-finger prototypes more recently introduced to market, two EMG inputs are used to simultaneously open and close all the fingers.

These multi-finger hands have multiple motors to control different fingers and different pre-programmed hand positions that the user can select from. Once the hand position is selected, the user can use myoelectric signals to control the opening and closing of the hand [14]. Multi-articulating hands include: the Michelangelo from Otto Bock,[1] iLimb-ultra from Touch Bionics[2] and BeBionic V3 from RSLSteeper,[3] shown in Fig. 7.21.

In general, prosthetic devices that enable movements with several degrees of freedom perform a sequential control and implement locking mechanisms or signals interruption to change from one degree of freedom to another. In contrast, the human neuro-muscular system smoothly and simultaneously articulates multiple degrees of freedom. Therefore, users often feel that the control of their prosthesis is not intuitive, generating often long disappointing periods of training and learning [15].

Currently prosthesis rejection is driving recent advances in mechanical prostheses to develop devices with a functionality comparable to an intact human hand. Clear examples are the SmartHand [16], the VU hand [17], and the DARPA RP 2009 Intrinsic Hand [18], which has 18 degrees of freedom. These prostheses have not been yet used in clinical practice due to the lack of adequate control algorithms.

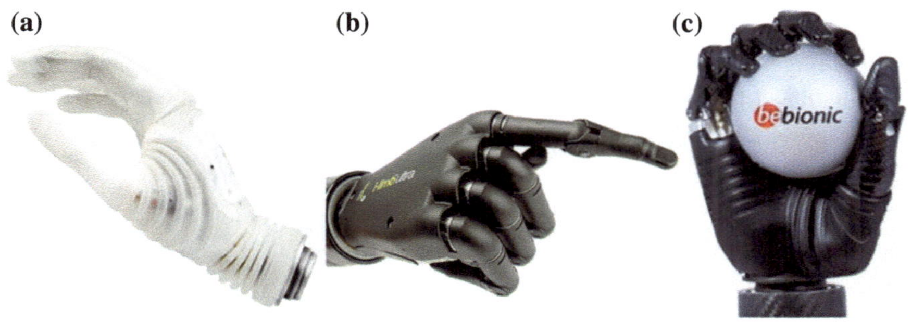

Fig. 7.21 Most advanced multi-articulating hands currently available. **a** Michelangelo, **b** i-limb-ultra-gallery3, **c** bebionic3

[1] http://www.ottobock.com/cps/rde/xchg/ob_com_en/hs.xsl/49464.html

[2] http://www.touchbionics.com/products/active-prostheses/i-limb-ultra/

[3] http://bebionic.com/

In this section, we are going to use one of the latest developed prosthesis by Otto Bock, the Michelangelo hand, to show an application of using the grasp quality measures studied in the previous chapter to evaluate prosthesis grasps and compare them with the values obtained for the human hand.

7.8.1 Material and Methods

7.8.1.1 The Michelangelo Hand

The Michelangelo hand is a five-finger prosthetic hand which has the ability to separately position the thumb using muscle signals, which allows seven different hand positions (see Fig. 7.22).

Additionally, it includes a flexible wrist joint which permits flexion, extension and rotation. A software and EMG signal processing utilize a graphic user interface, promoting control predictability. It has a natural-looking appearance with the digits covered in a soft silicone. The hand has a span of about 114 mm when the fingers are open. In addition, the oval wrist adapter looks more natural than a conventional prosthetic wrist.

The Michelangelo hand model for OpenRAVE without the cosmetic glove (Fig. 7.23) was provided by the *Vision4Robotics Group*[4] with permission from Otto Bock.

This prosthesis has three active fingers: the thumb, the index finger and the middle finger; the last two fingers are for cosmetic reasons. The index and middle fingers are mechanically coupled via a cable, so that usually they both come into contact with the object (as long as the difference in the joint angles at MCP joints of index and middle fingers is smaller than approx. 25°) and produce the same force. If the index finger come into contact with the object first, the middle finger still continue to move, until also the middle finger is in contact. Once the index and middle fingers stopped, the ring finger and little finger do not continue to move. If they come into contact with the object first, they flex, since they are coupled via soft plastic parts with the

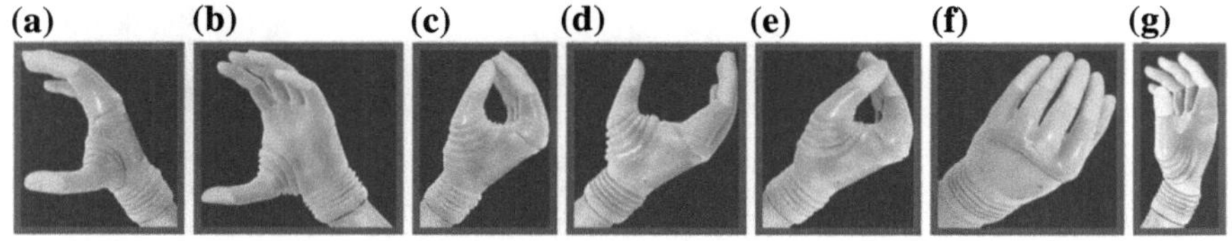

Fig. 7.22 Different grip types enabled by the Michelangelo hand changing the position of the thumb[5]: **a** open palm, **b** opposition power grip, **c** tripod pinch, **d** lateral power grip, **e** lateral pinch, **f** finger abduction/adduction and **g** neutral position

[4] Automation and Control Institute, Vienna University of Technology, Vienna, Austria.

main drive of the Michelangelo hand. This behaviour was mimicked implementing a closure algorithm for the precision grasp. Figure 7.24 shows the prosthetic hand model most open and closed postures.[5]

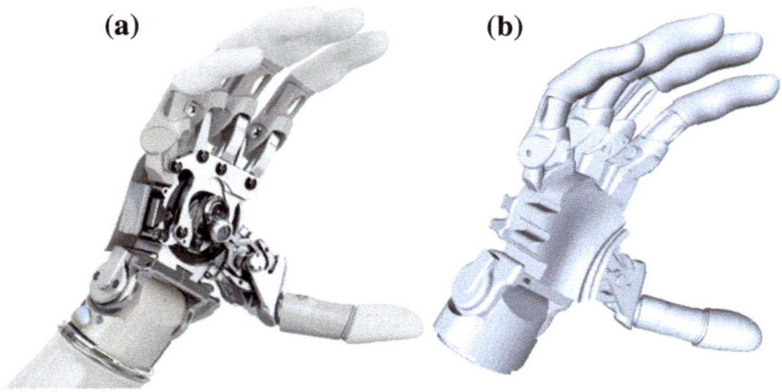

Fig. 7.23 Otto bock Michelangelo hand: real (Photo courtesy of Otto bock) and model implemented in OpenRAVE. **a** Real **b** simulated model

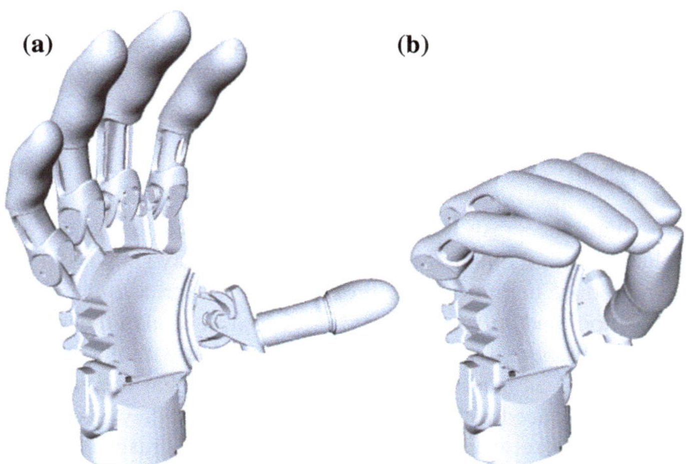

Fig. 7.24 Most open and closed postures using the closure algorithm of the Michelangelo hand using the *tripod pinch* thumb posture

[5] http://www.living-with-michelangelo.com/

7.8.1.2 Grasp Quality Measures Applied to Hand Prosthesis

The quality measures selected to evaluate the independent aspects of the human hand grasp were considered to evaluate the grasp of this hand prosthesis. However the meaning of some of this aspects vary or are not applicable to a prosthesis.

The first three aspects: **Restriction of the grip** (Q_{A3_N}), **Ability to resist forces** (Q_{C1_N}) and **Dynamic effects** (Q_{B1_N}) are metrics adopted from robotics which use the contact points to evaluate the grasp, therefore they can be used for the Michelangelo hand in the same way.

The aspect of **Comfort** measured by Q_{D1_N} does not refer in this case to the discomfort produced by stresses of the articular soft tissues when they come stretched close to the joints operating limits, but now must be understood as a reduction of manipulability, as possible changes in the grasp posture are limited.

The **Manipulability** of the prothesic hand, measured by Q_{D2_N}, is always zero, given that the hand has only one degree of freedom and therefore one of the singular values of the grasp Jacobian matrix (G_J) is always zero. This shows that, once achieving the grasp posture, the hand can not produce other movements to the object, which have to be produced by the wrist or arm, producing a manipulability of the hand equal to zero.

The **Muscular fatigue** has no meaning for a prosthesis without muscles and therefore it was discarded.

In order to make the values of these measures comparable with the ones obtained using the human hand model, we normalized them using the same ranges previously proposed (see Table 7.6).

7.8.1.3 Hand postures

In order to evaluate the grasp quality of the Michelangelo hand, we used the 36 postures evaluated with the human hand model (previously shown in Fig. 7.3) and attempted to reproduce them using the prosthesis. However, as this hand has predefined thumb positions and only one degree of freedom, several of the grasps achieved with the human hand were not possible to be reproduced with the prosthesis. Claw grasps were discarded as the hand has fixed interphalangeal joints. Lateral grasps could have been reproduced with the *lateral power grip*, but they were also discarded as this predefined posture would result in only two-finger grasps (thumb and index fingers). Therefore, only the 12 cylindrical grasps were considered using the *tripod pinch* posture shown in Fig. 7.22c.

In order to reproduce the same postures, the wrist was located in the same position registered for the human hand using the VICON motion capture system (described in Sect. 5.6.2). At this position, the closure algorithm was executed until all fingers have contacted the object or reached their maximum limits. The collision algorithm is used to determine the contact points and normals and all the selected quality measures were evaluated. All three-finger grasps were performed closing all fingers as the Michelangelo hand can not control the fingers used to perform the grasp. However, as only the thumb, index and middle are active fingers, they are the only ones that produce contact forces. The ring and the small finger produce only a small force when grasping an object, since they are coupled using springs with the main drive of the Michelangelo hand. The main reason for the ring and small finger is cosmetics, nevertheless—also with small grip forces—these fingers can help to stabilise an object, depending on their shape.

Therefore, when trying to simulate three-finger grasps, grasps with 3, 4 or 5 fingers were achieved, depending if the ring and small fingers reached the object. But only the three fingers selected to perform the grasp produced active forces. On the other hand, when trying to simulate five-finger grasps, sometimes the prosthesis achieved grasps with four fingers, and again, only three of the contacting fingers produced active grasping forces. We have—in both cases—calculated two values for each quality measure: (i) one only using the three active fingers and (ii) using the contacts actually produced; and present these two values as a range among which the grasp quality measure actually lies.

7.8.2 Results

The results of calculating the independent grasp aspects for the 12 cylindrical postures are presented in Table 7.12 taking into account both: only the 3 active fingers and the real number of contacts. The values for each range did not vary significantly for the different aspects. The *comfort* was not altered given that, as the fingers move according to one degree of freedom, all the fingers were at the same position in the joint range of motion. The *restriction of the grip* varied less than 1 % of the total normalized range [0–1]. The *dynamic effects* aspect presents variations (up to 15 %) as the centroid of the grasp polygon varied between three and five fingers, reducing or increasing its distance to the object center of mass. The *ability to resist forces* was the one with greater variation (up to 19 %) given that if the number of fingers increases, the ability of a grasp to resist external forces also increases, and viceversa.

In order to better visualize these results, each grasp is depicted in Fig. 7.25. They are presented alongside the picture of the respective human hand grasp and radar plots present the selected 5 independent aspects evaluated for each grasp and both hands.

The values obtained measuring the *restriction of the grip* are similar for both hands, only giving better values for the human hand in the horizontal three-finger posture grasping the cylinder in the center.

The *ability to resist forces* gave equal or better results in most of the cases (5 over 6) for the grasp performed with the Michelangelo hand for three finger grasps. In these cases, the grasp was performed with three fingers with the human hand for which the Michelangelo hand used 5, allowing it to increase its ability to resist forces. It can be visualized that when the measure is evaluated for the 3 active fingers of the Michelangelo hand, it is equal or less than for the human hand. In contrast, it gave generally (5 over 6) better results for the grasp performed with the human hand for

Table 7.12 Results of the grasp independent aspects for the 12 cylindrical postures using the Michelangelo hand using the 50 mm diameter cylinder

Posture			F[a]	Rest. of the grip Q_{A3_N}		Ability to resist f. Q_{C1_N}		Dynamic effects Q_{B1_N}		Comfort Q_{D1_N}	
				3 active fingers	Real contacts	3 active fingers	Real contacts	3 active fingers	Real contacts	3 active fingers	Real contacts
Horizontal	Left	3 [a]	5	0.0743	0.0687	0.1889	0.2567	0.4281	0.5365	0.7893	0.7893
		5	4	0.0743	0.0687	0.2033	0.2900	0.4416	0.4948	0.7712	0.7712
	Center	3	4	0.0717	0.0650	0.0189	0.1700	0.7884	0.8138	0.7326	0.7326
		5	4	0.0753	0.0687	0.1133	0.2800	0.8642	0.8676	0.7226	0.7226
	Right	3	4	0.0740	0.0673	0.0889	0.2289	0.3616	0.2956	0.7828	0.7828
		5	5	0.0757	0.0677	0.1989	0.2167	0.5498	0.4061	0.8307	0.8307
	Up	3	3	0.0743	0.0743	0.1411	0.1411	0.1790	0.1790	0.8242	0.8242
Vertical		5	5	0.0740	0.0650	0.0600	0.2367	0.1632	0.2823	0.8305	0.8305
	Center	3	4	0.0743	0.0680	0.0511	0.1956	0.8482	0.8549	0.7949	0.7949
		5	4	0.0720	0.0660	0.0378	0.2189	0.7554	0.7866	0.7623	0.7623
	Down	3	3	0.0740	0.0740	0.0689	0.0689	0.3491	0.3491	0.8170	0.8170
		5	4	0.0753	0.0687	0.1311	0.3078	0.5435	0.4718	0.7900	0.7900

[a] Num fingers that were planned
[b] Num fingers that actually contacted the object

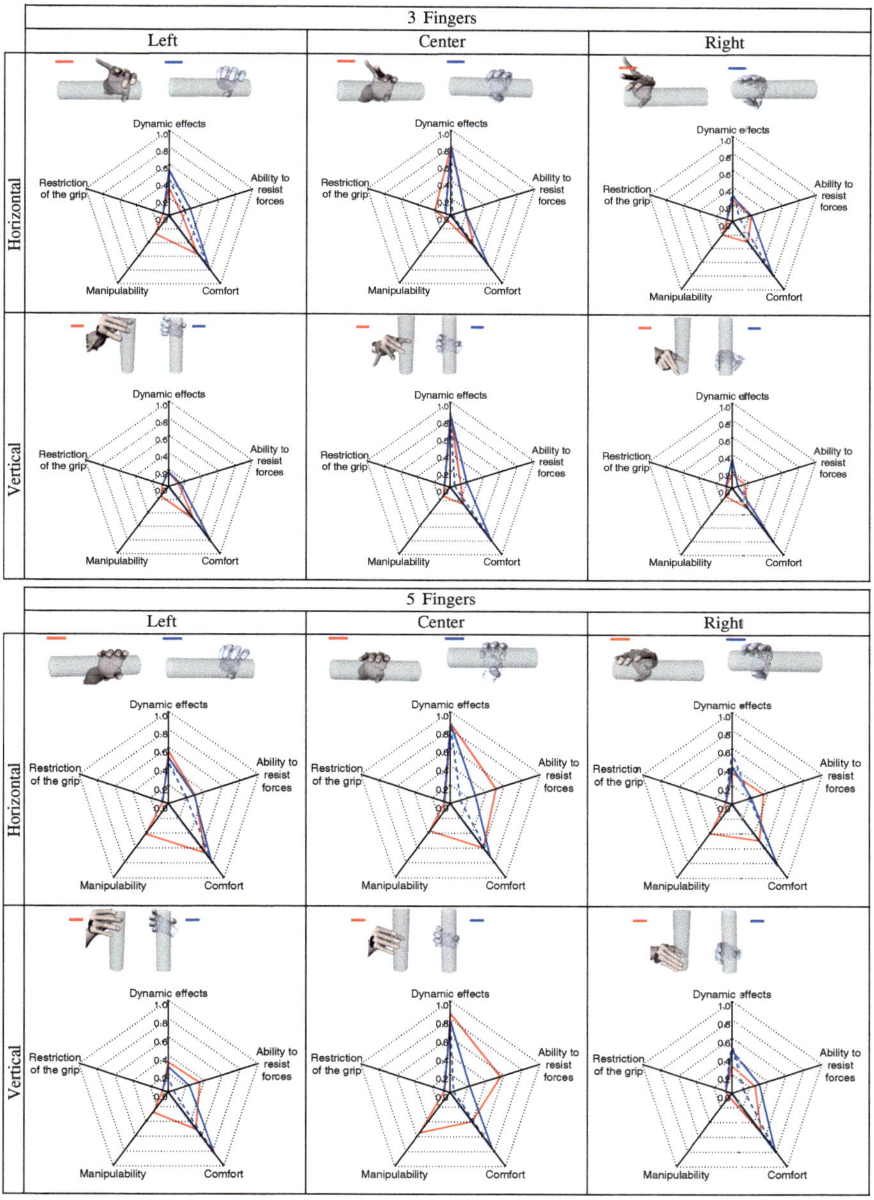

Fig. 7.25 Independent grasp aspects for the human and Michelangelo hands (red line: Human, continuous blue line: Michelangelo calculated with all fingers contacted and dotted blue line: Michelangelo calculated with only the 3 active fingers)

five finger grasps, given that in many of these cases only 4 fingers of the Michelangelo hand actually contacted the object.

The *dynamic effects*, measured with the distance from the grasp polygon to the center of mass, gave similar results for both hands since the wrists were located in the same position. The results clearly show that grasps performed close to the object centre are rated better than in the extremes. In some cases, there are small differences given that the human or Michelangelo hand positioned the fingers differently.

The grasp *comfort* is measured taking into account how far are the hand joints from their limits. As it was mentioned before, it should be understood as a measure of manipulability in terms of the hand ability to move the joints after performing the grasp. Since the object had a medium size, the prosthesis is grasping it at approximately the center of its joints' range of motion thus getting always a very high performance in this aspect. This is more accentuated when the human hand is performing grasps with three fingers in which it moves the ring and small fingers to their limits to put them out of the way. In this cases, it should be considered to modify this measure to only take into account the joints' values of fingers that contact the object.

Lastly, the *Manipulability* is the aspect that clearly gives a large advantage to grips performed with the human hand since this measure always gives zero values for the prosthesis.

7.8.3 Discussion

In this section, we demonstrated one of the possible uses of the proposed framework for evaluating grasps through its various independent aspects. We showed how the grasp ability of a prosthetics hand can be compared to the one that the human hand has.

Although prosthetics hands have evolved and improved over the years, they have large deficiencies in their appearance and their functionalities are very limited. These deficiencies are largely related with the quality of the grasps they can perform.

Specifically, the Michelangelo hand largely lose dexterity and versatility compared to the human hand, reducing from 23 to only one degree of freedom. For this reason, out of 36 grasps that were studied for the human hand only 33 % were able to be evaluated for the prosthesis.

Additionally, the number of quality measures that are meaningful for the Michelangelo hand is reduced, excluding specially the proposed biomechanical indices as they do not make sense for hands without muscles.

The results given by the manipulability measure (Q_{D2}) are specially relevant for this hand. Although it was possible to calculate the singular values of the grasp Jacobian matrix, the inverse of the condition number always gave zero values. This demonstrates that once achieving the grasp posture, the hand can not produce other movements to the object, which have to be produced by the wrist or arm. This clearly shows one of the major deficiencies with the prosthetic hands: loss of manipulability.

As future work, additional measures might be proposed to take into account the fact that although the prosthesis is a five-finger hand, only three of its fingers are active and therefore able to produce significant forces to the grasped object. Specifically, the role of the passive forces generated by the ring and little fingers should be studied.

Additionally, future studies can investigate how adding different degrees of freedom to prosthetic hands can improve their ability to perform better grasps. This can be used by the robotics community since more and more robotic hands tend to be more similar to the human hand trying to achieve its dexterity to perform grasps and manipulate objects.

7.9 Conclusion

In this chapter, we studied the selected grasp quality measures to evaluate the human grasp. The fatigue and muscle safety margin indices were included to consider biomechanical aspects of the human hand. The varied nature of the measures results in very different dimensional units and ranges of variation. To make the indicators comparable, their mathematical limits—where they could be determined—were used to normalize them, so that they had the best value of 1 and the worst value of 0.

A variability analysis was performed changing different aspects that may influence the grasp such as the object's weight and size, the position, orientation or type of grasp or the number of fingers used. The results for the normalized indices showed that their values moved in different and small zones of the range 0–1. Therefore, the mathematical limits chosen did not seem to fit well to the real limits of variation of the measures of the grasping hand. Therefore new ranges were proposed to better normalize the measures. Additionally, a sensitivity analysis demonstrated that with a 10 % variation in the tentative grasp posture, most of the measures show less than 20 % variation over their proposed ranges, which shows their robustness to their input uncertainty errors.

Through a correlation analysis, groups of measures that evaluate similar aspects of the grasp were determined, allowing us to find a reduced number of indices to assess the overall quality of the grasp. A physical interpretation was given to the six independent aspects of evaluation: restriction of the grip, ability to resist forces, dynamic effects, comfort, manipulability and muscular fatigue.

These measures were evaluated using different grasps that were also reproduced by human subjects. A new generalized grasping quality index to represent all of the grasp indices as one unified measure is formulated by using a neural network. The human assessment was used to train the network and validate the results. As future work, the importance of each aspect being measured by the indices could be modified to have into account the manipulation task to be performed after the grasp.

The proposed framework for human hand grasping has several applications but in this chapter we demonstrated how it can be applied to improve the design of hand prosthesis or robotic hands, comparing the results of the grasp quality measures with the ones obtained for the human hand model.

The experiments performed in this study are shown as an example of the procedure that could be applied to evaluate the human grasp. However, the results can be extended by performing further experiments varying more aspects of the grasp. These include using objects of different shapes, increasing the variation of their weight, as well as varying the task to perform while grasping. In addition, a quantitative assessment instead of a ranking of grasps would allow us to obtain a global measure that can then be more easily compared with the results of the quality indices. The subjects could also be asked to assess the different characteristics of the grasp which are somehow related with the results of the selected measures to gain additional information.

All grips studied in this work are precision grasps, in which only the distal phalanges of the fingers are in contact with the objects. Therefore, future studies can also examine how to adapt—and even propose—quality measures for power grasps where other segments of the fingers and the palm may also be in contact with objects.

References

1. Boivin, E., Sharf, I., Doyon, M.: Optimum grasp of planar and revolute objects with gripper geometry constraints. In: Proceedings of the IEEE International Conference on Robotics and Automation, vol. 1, pp. 326–332 (Apr–May 2004).
2. Chinellato, E., Fisher, R., Morales, A., del Pobil, A.: Ranking planar grasp configurations for a three-finger hand. In: Proceedings of the IEEE International Conference on Robotics and Automation, vol. 1, pp. 1133–1138 (Sep 2003).
3. Chinellato, E., Morales, A., Fisher, R., del Pobil, A.: Visual quality measures for characterizing planar robot grasps. IEEE Trans. Syst. Man Cybern. Part C Appl. Rev. **35**(1), 30–41 (2005)
4. Kim, B.H., Yi, B.J., Oh, S.R., Suh, I.H.: Non-dimensionalized performance indices based optimal grasping for multi-fingered hands. Mechatronics **14**(3), 255–280 (2004)
5. Vergara, M., Pérez-González, A., Serrano-Cabedo, J., Rodríguez-Cervantes, P.: Resultados de un trabajo de campo sobre agarres utilizados en tareas cotidianas. In: XIX Congreso Nacional de Ingeniería Mecánica (CNIM 2012), Castellon, Spain (Nov 2012).
6. Sancho-Bru, J., Vergara, M., J.B.N.J., Mora Aguilar, M., Pérez-González, A.: Medición del movimiento de todos los segmentos de la mano mediante videogrametría. In: XIX Congreso Nacional de Ingeniería Mecánica (CNIM 2012), Castellon, Spain (Nov 2012).
7. Darling, W.G., Cole, K.J., Miller, G.F.: Coordination of index finger movements. J. Biomech. **27**(4), 479–491 (1994), http://www.ncbi.nlm.nih.gov/pubmed/8188728
8. Rash, G.S., Belliappa, P.P., Wachowiak, M.P., Somia, N.N., Gupta, A.: A demonstration of validity of 3-d video motion analysis method for measuring finger flexion and extension. J. Biomech. **32**(12), 1337–1341 (1999), http://www.ncbi.nlm.nih.gov/pubmed?term=A%20demonstration%20of%20the%20validity%20of%20a%203-D%20video%20motion%20analysis%20method%20for%20measuring%20finger%20flexion%20and%20extension
9. Yun, M.H., Freivalds, A.: Analysis of tool Grip tasks using a 3-D glove. IOS Press, Michigan, USA (1998)
10. Speirs, A.D., Small, C.F., Bryant, J.T., Pichora, D.R., Zee, B.Y.: Three-dimensional metacarpophalangeal joint kinematics using two markers on the phalanx. Proc. Inst. Mech. Eng. H **215**(4), 415–419 (2001), http://www.ncbi.nlm.nih.gov/pubmed/11521764

11. Vergara, M., Sancho-Bru, J.L., Pérez-González, A.: Description and validation of a non-invasive technique to measure the posture of all hand segments. J. Biomech. Eng. **125**(6), 917–922 (2003), http://www.ncbi.nlm.nih.gov/pubmed?term=A%20demonstration%20of%20the%20validity%20of%20a%203-D%20video%20motion%20analysis%20method%20for%20measuring%20finger%20flexion%20and%20extension
12. Hudgins, B., Parker, P., Scott, R.: Control of artificial limbs using myoelectric pattern recognition. Med. Life Sci. Eng. **13**, 21–38 (1994)
13. Merrill, D.R., Lockhart, J., Troyk, P.R., Weir, R.F., Hankin, D.L.: Development of an implantable myoelectric sensor for advanced prosthesis control. Artif. Organs **35**(3), 249–252 (2011), http://dx.doi.org/10.1111/j.1525-1594.2011.01219.x
14. Harvey, Z.T., Potter, B.K., Vandersea, J., Wolf, E.: Prosthetic advances. J. Surg. Orthop. Adv. **21**(1), 58–64 (2012), http://www.ncbi.nlm.nih.gov/pubmed/22381512
15. Engeberg, E.D.: A physiological basis for control of a prosthetic hand. Biomed. Signal Process. Control **8**(1), 6–15 (2013), http://www.sciencedirect.com/science/article/pii/S1746809412000717
16. Cipriani, C., Controzzi, M., Carrozza, M.C.: The smarthand transradial prosthesis. J. Neuroeng. Rehabil. Signal Process. **8**, 29–29 (2011), http://www.ncbi.nlm.nih.gov/pubmed/21600048?dopt=Abstract&holding=f1000, f1000m, isrctn
17. Dalley, S., Wiste, T., Withrow, T., Goldfarb, M.: Design of a multifunctional anthropomorphic prosthetic hand with extrinsic actuation. IEEE/ASME Trans. Mechatron. **14**(6), 699–706 (Dec 2009).
18. Weir, R., Clark, S., Mitchell, M., Puchhammer, G., Kelley, K., Haslinger, M., Kumar, N., Hofbauer, R., Kuschnigg, P., Cornelius, V., Eder, M., Grausenburger, R.: New multifunctional prosthetic arm and hand systems. In: Engineering in Medicine and Biology Society, 2007. EMBS 2007. 29th Annual International Conference of the IEEE. pp. 4359–4360 (Aug 2007).

Chapter 8
Conclusions

Grasping is one of the most challenging problems in robotics, and require knowledge and development from different fields. The problem, like many others in robotics, benefits greatly from the use of a simulator. First of all, the simulation might be used to replace the real hardware to the extent that it is capable of reproducing the actual physical behaviour, which is of special importance in the context of robot manipulation. Second, it might be used as a prediction engine that can help to understand the effects of actions and provide the base for developmental learning. Additionally, if simulation accurately reproduces the real sensor and actuator feedback, robots might automatically learn from low-level sensor inputs without the depreciation of real hardware.

The quality of simulations has improved rapidly in recent years with the development of different physics engines and a wide variety of 3D rendering engines. This has enabled simulators to improve their prediction capabilities. As a result, several approaches have been proposed to use simulation to test different grasp hypotheses, select a stable grasp using different criteria and plan to execute and monitor the complete grasping task. However, performing dexterous manipulation in complex environments is still a very challenging task in robotics.

In contrast, humans have the ability to perform a great variety of complex tasks in a very dexterous way. Therefore, the development of models and algorithms able to replicate the human grasp could help us in understanding how the human brain is able to plan and execute these complex actions and improve our robots by providing them with more dexterous manipulation abilities.

In this book, we tackled the problem of grasping simulation using an interdisciplinary approach. On the one hand, robot grasping simulation has been improved using state-of-the-art tools to better predict the reality. On the other, a biomechanical model able to represent the human hand has been developed. The robotic and biomechanical knowledge was used to propose different quality measures to evaluate different aspects of the human grasps. Finally, the knowledge acquired from the evaluation of grasping in humans is compared with grasping performed by a prosthetic hand showing how the gap between robot and human grasp manipulation could be reduced.

B. León et al., *From Robot to Human Grasping Simulation*,
Cognitive Systems Monographs 19, DOI: 10.1007/978-3-319-01833-1_8,

The proposed framework for human hand grasping has several applications in various fields of research related with manipulation. They include the development of more efficient robotic manipulators for replacing the human hand assembly lines in industry thus freeing operators from repetitive tasks; the development of robotic arms for assisting people with disabilities; the improvement of the design of hand prosthesis or robotic hands; the development of devices that transmit haptic touch sensation during the manipulation processes in virtual reality tele-operation or robotic tasks, such as during tele-surgery; the planning of medically-oriented operations to restore or enhance the capabilities of patients with hand grasp pathologies; or the design of hand tools and other products used by human hands.

Although dexterous manipulation remains as one of the biggest challenges in robotics, this book presents approaches that aim to provide a procedure that opens the way to achieve this goal.

This chapter highlights the main contributions presented in this manuscript and provided some ideas that can be used in future works to extend the approach and applications presented here.

8.1 Contributions

The specific contributions of this book can be detailed and subdivided following its two main parts: robot and human grasping simulation.

8.1.1 Robot Grasping Simulation

8.1.1.1 OpenGRASP

This book has presented a fully operational simulation toolkit for robot grasping and manipulation. Its main design principles are extensibility, interoperability and public availability. In its development we have used existing and widely-available components to ensure its standardization and easy adoption. We have also provided additional tools and features that provide users with a quick start: the robot editor based on Blender, COLLADA file format, two Physics Abstraction Layers (PAL and FISICAS), and models of existing robot hands [1, 2].

8.1.1.2 Tactile Sensor Model

This book has addressed the problem of creating a simulation of a tactile sensor as well as its implementation in a simulation environment. The simulated tactile sensor model utilizes collision detection and response methods using soft contacts as well as a full friction description. The tactile element is created based on a geometry enabling the creation of a variety of different shape tactile sensors. The tactile sensor element can be used to detect touch against triangularized geometries. This independence in shape enables the use of the sensor model for various applications, ranging from regular tactile sensors to more complex geometries as the human hand which makes it possible to explore human-like touch. The developed tactile sensor model is implemented within OpenGRASP and is available as an open-source plugin. The model has been validated through several experiments ranging from physical properties verification to testing on robot grasping applications. This simulated sensor can provide researchers with a valuable tool for robotic grasping research, especially in cases where the real sensors are not yet accurate enough [3–5].

8.1.1.3 Grasping Known Objects Using a Model-Driven Grasping Approach

We have presented the Model-Object Overlap Metric to incorporate information obtained from the depth-sensor into the grasp planning process. We have shown how this metric can be incorporated at different stages and in different fashions to conventional grasping pipelines to increase efficiency and robustness to errors from recognition. The simulator was used to find stable grasps ranked according to the overlap quality metric in relation to the observed data.

8.1.1.4 Grasping Unknown Objects

The applicability of simulation to support the execution in the real-world in a service robotic scenario in which the graspable objects were unknown to the robot has been demonstrated and their shapes were predicted using symmetries. The simulator was used to select stable grasps and plan collision-free movements to be executed by the robot [6].

8.1.1.5 Grasping Familiar Objects According to Their Category and Given Task

We presented a grasping pipeline that allows autonomous robot grasping according to an object's category and a given task. Several state-of-the-art modules performing scene exploration through gaze shifts, segmentation, object categorization and task-based grasp selection were integrated. We showed how this allows the robot to transfer task-specific grasp experience between objects of the same category. The effectiveness of the approach is demonstrated on two humanoid robots: ARMAR-IIIa and Tombatossals. The simulator was used to generate off-line a set of task-ranked grasp hypotheses. This involves the generation of grasp hypotheses and their ranking according to the task and the object's category. In the on-line process, the most similar object model and the given task serve as a look-up in this database to retrieve the highest-ranked grasp [7].

8.1.1.6 Dynamic Grasping Simulation

We addressed the problem of simulating the complete dynamics of a humanoid torso robot, which includes sensors and actuators, in the context of robot grasping and manipulation. We evaluated the extent to which the simulation resembles the real behavior of manipulation tasks by using the same controller on the real and the simulated platforms and analyzing the differences. The results have shown that it is possible to simulate manipulation tasks with the current state of the art of simulation tools. Although the precision is not perfect, the framework is able to perform manipulation tasks using the same controller as that used in the real world with very similar results. An important result is that with this framework, it is possible to use simulated tactile data in manipulation task controllers that use tactile feedback [8, 9].

8.1.2 Human Grasping Simulation

8.1.2.1 Human Hand Model

A realistic and self-contained biomechanical model of the hand has been developed by merging the current knowledge of biomechanics, ergonomics and robotics. The model simulates the complete hand and can be easily scaled to study different percentiles of populations. It has a realistic representation that allows the ergonomic evaluation of products. The model is dynamic and can be used to study the muscular patterns associated with a specific grasp. It allows predicting feasible grasping postures and provides the contact information required for evaluating the grasp [10–12].

8.1.2.2 Human Grasping Simulation

The biomechanical simulation of the human hand grasping has been performed by adding to the biomechanical model the dynamic equilibrium of the grasped object. Different objective functions have been studied to solve the indeterminate problem of finding the muscular forces enabling the grasp. The main contribution is in showing that the widespread *Crowninshield minimisation function* does not work well when trying to simulate the grasping of an object with an already validated 3D model of the hand, and that the consideration of a stability criterion improves the estimations. This result is significant in the context of human grasp modelling and has not been reported previously in the literature, contributing to a better understanding of the human grasp [13].

8.1.2.3 Human Grasp Evaluation

Based on a review of robotic grasp quality measures, the most common measures have been adapted to the evaluation of the human grasp. Furthermore, two new measures, the *fatigue index* and *the muscle safety margin index* have been proposed to consider the biomechanical aspects of the human hand not taken into account by the existing robotic measures. The Monte Carlo method has been used to successfully obtain the upper limit for the robotic measure Q_{C2} in its adaptation of human grasp evaluation in order to illustrate its potential for the normalization of any of the measures [14, 15].

8.1.2.4 Variability and Sensitivity of the Quality Measures

The selected measures were used to evaluate the quality of different grasps. A variability analysis was performed changing different aspects that may influence the grasp such as the object's weight and size, the position, orientation or type of grasp or the number of fingers used. With these results, ranges to better normalize the measures were proposed. The robustness of the measures to uncertainty in their input data has been demonstrated through a sensitivity analysis, by randomly varying the tentative grasping posture within a 10 % range of variation [16].

8.1.2.5 Independent Grasp Aspects

Through a correlation analysis, groups of measures that evaluate similar aspects of the grasp were determined, allowing us to find a reduced number of indices to assess the overall quality of the grasp. A physical interpretation was given to the six independent aspects of evaluation: restriction of the grip, ability to resist forces, dynamic effects, comfort, manipulability and muscular fatigue [15].

8.1.2.6 Global Grasp Quality Index

These measures were evaluated for a set of different grasps that were also experimentally assessed by human subjects. The use of a neural network is proposed to merge all representative quality measures of the independent aspects, generating a generalized grasping quality index that provides a global assessment of the grasp. The human assessment was used to train the network and validate the results.

8.1.2.7 Application to Prosthetic Hand Grasping Evaluation

We have shown an application to use the developed human grasping simulation framework. The independent grasp quality measures were used to evaluate grasps produced by the Michelangelo prosthetic hand and compare them with the human results. The *muscular fatigue* was discarded given that it can not be applied for the prosthetic hand and the measure of *manipulability* always gave zero results for this one-degree of freedom hand. As prosthesis functionality becomes more similar to an intact human hand, more aspects can be used to evaluate its capabilities. This will give insights on how to improve prosthesis design to achieve a similar gripping ability to the human hand, or even improve it.

8.2 Future Work

The approaches proposed in this book solve some problems related to grasping simulation and evaluation. However, much work is still required in these fields. The following are some open research lines that can be further studied.

8.2.1 Tactile Sensor Model

The collision detection method for solving the contact points is currently a brute force method. Improvements on this area can greatly improve the overall computation times of the tactile sensor model. Future work includes improving the collision detection times and improving computational efficiency.

8.2.2 Dynamic Grasping Simulation

The main drawback of the presented approach is the duration that the simulator requires to perform the same task as the real robot does, limiting its possible use as a prediction engine. However, recent advances in parallelizing processes to improve the speed of simulations can be a feasible solution to this issue. The parallel quick-step package available in ROS, provides an implementation in CUDA, OpenCL and OpenMP which accelerates the process of calculating each time step. Another challenging problem is the parameter setting for the simulator, for which learning algorithms to find the right parameters could be used.

8.2.3 Human Hand Model

There are several parts of the model that are open for improvement. Modifications of the thumb kinematics taking into account recent biomechanical studies [17, 18] could improve the model's estimation of the contact force producing by the thumb.

Additionally, the skin model can be improved at the joint zones by considering a deformation algorithm depending on the joint angles. That would enable more realistic hand postures. Additionally, neural networks can be used to avoid the experimental measure of the most open and tentative grasp postures. Finally, the graphical interface could also be improved to provide more options to the user.

8.2.4 Central Nervous System Criterion

Although the stability measure provided good results in these simulations, it may fail for others entailing certain level of manipulability, as the criterion selected by the CNS in each case will probably be a function of the task to be performed. Further research on the application of other robotics grasp quality measures for different tasks involving different levels of stability and manipulability is needed. Also, the model presented in this study has been used to study only grasps of cylinders with the fingertips. More complex grasps, involving more contact zones and more complex object geometries should be investigated in the future.

8.2.5 Human Hand Evaluation

More research is needed to investigate other biomechanical measures that might be obtained from the use of existing biomechanical hand models. Additionally, measures that consider the task to be performed while grasping could be included in the evaluation as it is an important aspect. The measures used to evaluate the grasp take only the hand into consideration. Including the whole arm into the evaluation—specially when evaluating the muscular fatigue—could improve the ability to choose more comfortable grasps.

8.2.6 Global Grasp Quality Index

The experiments performed in this study are shown as an example of the procedure that could be applied to evaluate the human grasp. However, the results can be extended by performing further experiments varying more aspects of the grasp. These include using objects of different shapes, increasing the variation of their weight, as well as varying the task to perform while grasping. In addition, a quantitative assessment instead of a ranking of grasps would allow us to obtain a global measure that can then be more easily compared with the results of the quality indices. The subjects could also be asked to assess the different characteristics of the grasp which are somehow related with the results of the selected measures to gain additional information.

References

1. León, B., Ulbrich, S., Diankov, R., Puche, G., Przybylski, M., Morales, A., Asfour, T., Moisio, S., Bohg, J., Kuffner, J., Dillmann, R.: OpenGRASP: a toolkit for robot grasping simulation. In: Simulation, modeling, and programming for autonomous robots, lecture notes in computer science, vol. 6472, pp. 109–120. Springer, Berlin (2010)
2. León, B., Puche, G., Martí, H., Morales, A.: A toolkit for robot grasping simulation. In: Robot 2011. Sevilla, Spain (2011)
3. Moisio, S., León, B., Korkealaakso, P., Morales, A.: Simulation of tactile sensors using soft contacts for robot grasping applications. In: Robotics science and systems 2011. Worshop on toward high-performance computing support for the analysis, simulation, and planning of robotic contact tasks (2011)
4. Moisio, S., Leon, B., Korkealaakso, P., Morales, A.: Simulation of tactile sensors using soft contacts for robot grasping applications. In: 2012 IEEE International Conference on Robotics and Automation (ICRA). pp. 5037–5043. Saint Paul, Minnesota (May 2012)
5. Moisio, S., León, B., Korkealaakso, P., Morales, A.: Model of tactile sensors using soft contacts and its application in robot grasping simulation. Robot. Auton. Syst. **61**(1), 1–12 (2013), http://www.sciencedirect.com/science/article/pii/S0921889012001911
6. Bohg, J., Johnson-Roberson, M., Leon, B., Felip, J., Gratal, X., Bergstrom, N., Kragic, D., Morales, A.: Mind the gap - robotic grasping under incomplete observation. In: IEEE International Conference on Robotics and Automation (ICRA). pp. 686–693. Shanghai, China (May 2011), http://ieeexplore.ieee.org/xpls/abs_all.jsp?arnumber=5980354
7. Bohg, J., Welke, K., Leon, B., Do, M., Song, D., Wohlkinger, W., Madry, M., Aldoma, A., Przybylski, M., Asfour, T., Marti, H., Kragic, D., Morales, A., Vincze, M.: Task-based grasp adaptation on a humanoid robot. In: 10th International IFAC Symposium on Robot Control (SYROCO 2012). Dubrovnik, Croatia (Sep 2012)
8. León, B., Felip, J., Marti, H., Morales, A.: Simulation of robot dynamics for grasping and manipulation tasks. 12th IEEE-RAS International Conference on Humanoid Robots (Humanoids) (Nov 2012)
9. León, B., Felip, J., Martí, H., Morales, A.: Towards complete robot dynamic grasping simulation. In: 5th international conference on cognitive systems (CogSys) (2012), poster (2012)
10. Sancho-Bru, J., Mora, M., León, B., Pérez-González, A., Iserte, J.: Modelo de la mano para la evaluacion de agarres. In: Actas del X Congreso Iberoamericano de Ingeniería Mecánica, CIBIM10. pp. 2453–2462. Oporto, Portugal (2011)
11. Sancho-Bru, J.L., Antonio Pérez-González, Mora, M.C., León, B.E., Vergara, M., Iserte, J.L., Rodríguez-Cervantes, P.J., Morales, A.: Towards a realistic and self-contained biomechanical model of the hand. Theoretical biomechanics pp. 212–240 (2011), http://www.intechopen.com/articles/show/title/towards-a-realistic-and-self-contained-biomechanical-model-of-the-hand
12. León, B.E., Sancho-Bru, J., Morales, A., Pérez, A.: Modelo biomecánico de la mano orientado al agarre de objetos. In: Actas I Reunión del Capítulo Nacional de la Sociedad Europea de Biomecánica. Zaragoza, Spain (2011)
13. Sancho-Bru, J.L., Mora, M.C., León, B.E., Antonio Pérez-González, Iserte, J.L., Morales, A.: Grasp modelling with a biomechanical model of the hand. Computer Methods in Biomechanics and Biomedical Engineering pp. 1–14 (2012), http://www.tandfonline.com/doi/abs/10.1080/10255842.2012.682156
14. León, B., Sancho-Bru, J., Rodriguez, S., Morales, A., Pérez, A.: Robotic quality measures evaluating human grasp. In: 18th Congress of the European Society of Biomechanics (ESB 2012). Lisbon, Portugal (2012)
15. León, B.E., Sancho-Bru, J.L., Jarque-Bou, N.J., Morales, A., Roa, M.A.: Evaluation of human prehension using grasp quality measures. Int. J. Adv. Rob, Syst (2012)
16. León, B., Sancho-Bru, J., Morales, A., Pérez-González, A.: Indicadores de la calidad del agarre humano: rangos de variabilidad y análisis de sensibilidad. In: XIX Congreso Nacional de Ingeniería Mecánica (CNIM 2012). Castellon, Spain (Nov 2012)

17. Valero-Cuevas, F.J., Johanson, M.E., Towles, J.D.: Towards a realistic biomechanical model of the thumb: The choice of kinematic description may be more critical than the solution method or the variability/uncertainty of musculoskeletal parameters. J. Biomech. **36**(7), 1019–1030 (2003)
18. Wohlman, S.J., Murray, W.M.: Bridging the gap between cadaveric and in vivo experiments: A biomechanical model evaluating thumb-tip endpoint forces. J. Biomech. 46(5), 1014–1020 (2013), http://www.sciencedirect.com/science/article/pii/S0021929012006513

MIX
Papier aus verantwortungsvollen Quellen
Paper from responsible sources
FSC® C105338

If you have any concerns about our products,
you can contact us on
ProductSafety@springernature.com

In case Publisher is established outside the EU,
the EU authorized representative is:
Springer Nature Customer Service Center GmbH
Europaplatz 3, 69115 Heidelberg, Germany

Printed by Libri Plureos GmbH
in Hamburg, Germany